Anionic Surfactants-
Chemical Analysis

SURFACTANT SCIENCE SERIES

CONSULTING EDITORS

MARTIN J. SCHICK

Diamond Shamrock Corporation
Process Chemicals Division
Morristown, New Jersey

FREDERICK M. FOWKES

Chairman of the Department of Chemistry
Lehigh University
Bethlehem, Pennsylvania

OTHER VOLUMES IN PREPARATION

Anionic Surfactants - Chemical Analysis

edited by JOHN CROSS

Department of Chemistry
Darling Downs Institute of Advanced Education
Toowoomba, Queensland, Australia

MARCEL DEKKER, INC.   New York and Basel

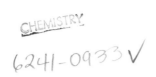

Library of Congress Cataloging in Publication Data

Main entry under title:

Anionic surfactants.

    (Surfactant science series ; v. 8)
    Includes bibliographical references and indexes.
    1. Surface active agents—Analysis. I. Cross,
John Thomas Daniel.

TP994.A58   668'.1   77-21835
ISBN 0-8247-6624-5

MARCEL DEKKER, INC.

270 Madison Avenue, New York, New York 10016

Current Printing (last digit):

10 9 8 7 6 5 4 3 2 1

PRINTED IN THE UNITED STATES OF AMERICA

PREFACE

Surface-active agents (surfactants) are currently manufactured in quantities of the order of a million tons per annum, of which about 80% is of the
types broadly known as anionic surfactants. This book is part of a series
devoted to such compounds and presents a comprehensive coverage of all
aspects of analysis, from the measurement of trace quantities in water to
quality control during manufacture. It is aimed at readers with a general
background of chemistry that includes familiarity with the basic concepts of
classical and modern methods of analysis. Consequently, the introduction
to analytical techniques has been kept to a minimum. For example, the
chapter on gas chromatography contains no discussion on principles of
separation, detector choice, etc.; the reader is referred to monographs on
the particular technique or to the numerous general texts devoted to modern
instrumental methods of chemical analysis for such information. However,
no detailed familiarity with surfactants or surface activity is assumed.

The authors, representing an international collection of academic and
industrial institutions, have been encouraged to present a critical, up-to-date
review of their topic as they themselves see it. In keeping with previous
volumes in the series, no attempt has been made to present a group opinion,
nor to inhibit the authors by confining them to a fixed format of presentation
or a specified range of common surfactants. Each contribution represents
a full treatise on a particular aspect of the topic in its own right and consequently readers should find themselves able to use any chapter independently
of the remainder. In addition to the material presented in this volume, a
chapter on the analysis of raw materials and formulations, automation, and
on-stream methods was planned, but unfortunately it was not possible to
arrange for coverage of these areas.

Two particular points concerning the contents of this book merit mention in this preface. Firstly, as conceived by Swisher in Volume 3 of this
series, the reference section at the end of each chapter contains the titles of
the articles in addition to the usual author and location information. Entry
to the text via the author index and further reference to the original articles

may thus be on a selective basis. The author index gives the page number on which a particular author's work is cited (even though this name may not necessarily be mentioned in the text), and the page on which the complete reference may be found.

Secondly, as authors, editors, publishers and readers alike are only too aware, research and development work continues during and beyond the time this book was in preparation. I have included in the introduction brief details of the services offered by some of the more prominent organizations dealing with standardization of analytical methods, so that in the years to come, the reader may be able to circumvent conventional literature searching and obtain details of the latest accepted methods from these agencies. This list was not intended to be exhaustive and the omission of reference to a particular organization should not be regarded as indicative that this author considers their work to be of little merit.

My sincere thanks are due to the many individuals and organizations who have supplied me with helpful information, to the management of the Darling Downs Institute of Advanced Education for the availability of facilities and their encouragement to undertake this task, to the authors of each contribution and their companies and finally, but not least, to my wife for her patience and aid in the preparation of the manuscript.

John Cross

CONTENTS

LIST OF CONTRIBUTORS

JOHN CROSS, Department of Chemistry, Darling Downs Institute of Advanced Education, Toowoomba, Queensland, Australia

DELIA M. GABRIEL, Unilever Research, Isleworth Laboratory, Isleworth, Middlesex, United Kingdom

ERICH HEINERTH,* Analytical Laboratories, Henkel and Cie, GmbH, Düsseldorf, Federal Republic of Germany

HANS KÖNIG, Analytical Laboratories, Blendax-Werke R. Schneider GmbH and Co., Mainz, Federal Republic of Germany

V. JOHN MULLEY, Unilever Research, Isleworth Laboratory, Isleworth, Middlesex, United Kingdom

TERUMICHI NAKAGAWA, Faculty of Pharmaceutical Sciences, Kyoto University, Kyoto, Japan

C. G. TAYLOR, Department of Chemistry, Liverpool Polytechnic, Liverpool, United Kingdom

TOYOZO UNO, Faculty of Pharmaceutical Sciences, Kyoto University, Kyoto, Japan

J. WATERS, Bioconsequences Section, Unilever Research Port Sunlight Laboratory, Port Sunlight, Wirral, Merseyside, United Kingdom

*Retired

vii

CONTENTS OF RELATED VOLUMES

INTRODUCTION

The term "surfactant" is a combination of "surface-active agent" and may be defined as any substance which is strongly adsorbed at a surface or interface. This definition encompasses the majority of substances which function as wetting, cleansing, and emulsifying agents and which are to be found wherever a stable solid-liquid suspension or liquid-liquid emulsion is formed, either during a cleansing process or in the preparation of consumer products such as cosmetics, toilet preparations, mayonnaise, and paints.

Surfactants comprise large molecules containing both nonpolar and polar (or ionic) groups which are referred to as the "hydrophobic" and "hydrophilic" sections, respectively. Traditionally they have been classified into anionic, cationic, nonionic, and ampholytic/zwitterionic types, according to the anture of the hydrophile. Works on the chemical analysis of nonionic and cationic surfactants have been presented elsewhere in this series [1, 2]. This volume is devoted solely to the analysis of anionic surfactants, which are commercially the most important, representing about 80% of the total surfactant production. Typical common examples are soaps [1], and salts of sulfated esters of fatty alcohols [2], and alkylbenzenesulfonates [3],

$$C_{17}H_{35}COO^-Na^+ \qquad C_{12}H_{25}OSO_3^-Na^+ \qquad C_{12}H_{25}C_6H_4SO_3^-Na^+$$

$$[\underline{1}] \qquad\qquad [\underline{2}] \qquad\qquad [\underline{3}]$$

but a quick survey of the patents in the Surface-active Agent Section of Chemical Abstracts in the last decade will show a formidable array of alternate choices. The chemistry of some of the more important members of this series has been dealt with in a previous volume [3].

As indicated above, anionic surfactants are to be found in a variety of consumer products and consequently in admixture with a wide variety of other substances. For example, in a formulated (or "built") laundry detergent, one would expect to find an anionic surfactant mixed with:

a. Sodium salts such as poly- and monophosphates, nitriloacetates, ethylenediaminetetraacetate, carbonates, and silicates, performing a variety of functions such as binding the surfactant into a free-pouring powder, controlling pH, increasing the ionic strength, preventing corrosion and complexing calcium and magnesium ions

b. Antideposition agents such as sodium carboxymethylcellulose

c. A fluorescent optical brightener

d. Sodium percarbonate or perborate (bleaching agent)

Alternatively the main ingredients of a toothpaste, in addition to a surfactant, are an abrasive (such as calcium carbonate or calcium hydrogen phosphate) slurried, for example, with a glycerol-sorbitol-water mixture to yield the typical viscous paste. Other components would be medicinal ingredients (antiseptics, fluorides, etc.) and flavoring compounds.

The pretreatments required to separate the surfactant from these two types of product will obviously differ considerably and when one considers the vast range of products which contain anionic surfactants, it will be appreciated that specific details for each individual case cannot be provided. Nevertheless, in Chapter 1 an exhaustive collection of methods has been condensed into an authoritative treatise on detection, separation, and isolation of surfactants.

Chapters 1 and 2 describe absorptiometric and some chromatographic methods for the identification of anionic surfactants. Perhaps the term "identification" is too definite, since most commercial products contain a mixture of homologues and isomers of the hydrophobic type, and full identification of each and every component is not always required.

Another system of surfactant classification which has become prominent in the last decade is based upon the susceptibility of the surfactant to degradation by living organisms found in natural waters. Surfactants can be classified as biodegradable (soft) or nonbiodegradable (hard) according to their ability to meet official specifications. A "soft" surfactant is generally one which is at least 80% decomposed within 21 days under specified test conditions; but these requirements may be tightened considerably in the not-too-distant future. Swisher [4], in an exhaustive treatise on surfactant biodegradation, has summarized the susceptibility of a surfactant to biodegradation as

a. Greatly influenced by the nature of the hydrophobe, being deterred by chain-branching in general and by the presence of a terminal t-butyl group in particular

b. Only slightly influenced by the nature of the hydrophile

c. Increased by increased distance between the hydrophile and the end of the hydrophobic chain

Analytical chemistry has played a vital role in the establishment of
these principles, and the development of chromatographic methods (Chapters 1 and 3), in particular, has permitted an accurate description of the
nature of the alkyl chains.

Chapter 4 describes the application of nuclear magnetic resonance
(NMR) spectroscopy to the examination of anionic surfactants. The author
has been the principal pioneer into this field that yields complementary information to the standard absorption spectroscopy measurements. A comprehensive collection of correlation tables, spectra, and commentaries
are presented.

The rapidly growing concern for the quality of surface waters, that developed as part of the general awakening in environmental matters, has
focused attention upon the need for methods for the determination of anionic
surfactants at the ppm level or lower. A colorimetric method was first
described in 1945 but the road to a technique that showed similar response
to a wide variety of anionic surfactants, was relatively unaffected by other
components of the sample, and gave consistent results in the hands of independent operators, has been long and arduous. Chapter 5 gives an up-to-
date account of the progress of these studies and also introduces some
newer, alternative methods. Many organizations are interested in improving
and standardizing procedures for the estimation of trace amounts of surfactants in water and also for the collection and preservation of samples. The
Water Pollution Control Federation (Washington, D.C.) has a membership
of 39 organizations within the United States and is affiliated with 24 over-
seas organizations; in addition to its regular journal it publishes an annual
literature review containing a section on detergents in which one can find
references to new developments in trace analysis, and also its well known
Standard Methods for Examination of Water and Waste Water (currently in
its 13th edition) in which the latest adopted procedures may be found. The
U.S. Environmental Protection Agency (EPA) (successor to the Federal
Water Pollution Control Administration) has adopted ASTM method D2330-68,
based upon the well-known methylene blue extraction technique.

When the concentration of surfactant in a sample can be better expressed
in percent than in ppm, the preferred method is volumetric over colorimetric
and the most popular methods are based upon the two-phase titration technique.
As in the case of colorimetry, a method has been available since the 1940s,
but one which satisfies the three criteria mentioned above has not been arrived at easily. In Chapter 6, Heinerth details the work of the Commission
Internationale d'Analyses (CIA), of which he was a member, and of the
Comité International des Dérivés Tensioactifs (CID), Paris, which has resulted in great improvements over the older methods.

Quality control of manufactured surfactants or surfactant-containing
formulations may require determination of a surprising number of common
impurities as well as the active ingredient itself. A simple soap, for example, may be tested for unsaponifiable matter, unsaponified matter,

unsaponified saponifiable matter, total fatty acid, free caustic alkali, free total alkali, chloride, moisture, volatile matter, alcohol-insoluble foreign matter, and glycerol content. Two organizations, in particular, have committees concerned with such problems: the International Standards Organization (ISO) and the American Society for Testing and Materials (ASTM). ISO comprises the national standard committees of 73 countries through which agencies its publications may be obtained. Its technical work is carried out by some 1,500 committees, subcommittees, and working groups, one of which, TC 91, is concerned with surfactants. The CIA submits its findings to ISO. Publications of ISO include an annual catalogue (with quarterly supplements), the 1974 version of which lists 33 accepted standards concerning surfactants; about half of these are analytical procedures. Each technical committee publishes an annual report referring to new and revised standards.

ASTM comprises many producer and consumer groups and likewise has a committee (D-12) on "Soaps and other Detergents" with a subcommittee (D-12-12) on "Analysis of Soaps and Synthetic Detergents" whose many task groups have arrived at adopted standards for sampling and analysis of soaps, soap products, and common ingredients of formulated detergents. Standards are published annually and a Committee D-12 Handbook was published in 1972 with the intention of annual updating.

No attempt has been made here to list all the types and trade names of anionic surfactants. Such a list would be rapidly out of date and would considerably increase size and cost. Throughout this text authors have identified the particular surfactants to which they have referred. If a more comprehensive list is required the reader is referred to the latest McCutcheon annual [5].

REFERENCES

1. M. Schick (ed.), Nonionic Surfactants, Marcel Dekker, New York, 1967.

2. E. Jungermann (ed.), Cationic Surfactants, Marcel Dekker, New York, 1970.

3. W. Linfield (ed.), Anionic Surfactants—Organic Chemistry, Marcel Dekker, New York, 1975.

4. R. Swisher, Surfactant Biodegradation, Marcel Dekker, New York, 1972.

5. Detergents and Emulsifiers, John W. McCutcheon, Inc., Morristown, N.J.

Anionic Surfactants-
Chemical Analysis

Chapter 1

DETECTION, SEPARATION, AND ISOLATION OF ANIONIC SURFACTANTS

Delia M. Gabriel and V. John Mulley

Unilever Research
Isleworth Laboratory
Isleworth, Middlesex, United Kingdom

I. CLASSIFICATION

Several classification schemes have been reported in the literature and most of them cover the whole range of surfactants, not just anionics. The underlying principle on which each classification is based is more or less related to the purpose of the classification (for example, analysis and product development). One of the earliest schemes is that of Brass and Beyrodt [1].

1

They investigated the problems associated with dyeing and printing, and suggested a classification for wetting agents by splitting them into aliphatic and aromatic compounds. The scheme was very simple and did not take ionic character into account nor did it cover the whole range of surfactants.

(1) Aliphatic compounds

    (a) Derivatives of hydroxy fatty acids and sulfo fatty acids (Turkey Red Oils)

    (b) Derivatives of fatty acids (Igepons)

    (c) Derivatives of fatty alcohols (Gardinols)

(2) Aromatic compounds

    (a) Derivatives of naphthalene (sulfonic acids of homologous naphthalenes)

    (b) Heterocyclic (pyridine) bases

Since this work was published in 1935 there has been a vast increase in the range of surfactants manufactured and today this scheme is quite inadequate.

In 1951, Balthazar [2] reported a scheme based on the source of the starting materials which divided the anionics into fatty acid and hydrocarbon derivatives as follows:

(A) Fatty acid derivatives

    (1) Sulfated fatty alcohols (Gardinol type)

    (2) Sulfonated amides (Igepon T type)

    (3) Sulfated amides (Deterpon type)

    (4) Sulfonated esters (Igepon AP type)

    (5) Sulfated esters (Halo, Syntex type)

    (6) Sulfonated esters of dicarboxylic acids (Aerosol type)

(B) Hydrocarbon derivatives

    (1) Alkylsulfonates (Mersolate type)

    (2) Alkylarylsulfonates

        (a) Short chain and a large cyclic (naphthalene) ring (Nelzal BX type)

        (b) Long chain and a small cyclic (benzene or toluene) ring (Nacconal type)

    (3) Secondary alkyl sulfates (Teepol type)

In 1957, Hintermaier [3] proposed a decimal classification consisting of up to six digits in two groups of three. The first digit represented one of his six main groups which were: 1, anionic; 2, cationic; 3, nonionic; 4, amphoteric; 5, neutral; and 6, mixtures. Subsequent digits indicated one or more surface-active groups in the molecule and the types of groups. For example, the coding for fatty acid soaps was 11 111. This type of scheme, however, requires a comprehensive code which must be updated whenever new types of surfactant are produced.

Sisley [4] based his scheme on the method of manufacture of the products and for anionic surfactants he lists 19 main groups, many of which are further subdivided. The main groups are as follows:

1. Anion-active substances

   1/A Products obtained by saponification of fatty substances

   1/B Products obtained by direct sulfonation of fatty substances without preliminary transformation

   1/C Products obtained by sulfonation of fatty acid esters

   1/D Products obtained by sulfonation of fatty acids with monovalent sulfonated alcohols

   1/E Sulfonated derivatives of fatty acid esters of low-molecular-weight (maleic, fumaric)

   1/F Products obtained by sulfonation of fatty amides

   1/G Products obtained by condensation of fatty acid chlorides with amines

   1/H Products obtained by sulfonation of fatty acid nitriles

   1/I Products obtained by sulfonation of fatty aldehydes

   1/J Products obtained by sulfonation of fatty ketones

   1/K Products obtained by sulfonation of natural and synthetic alcohols with six or more carbon atoms to produce sodium salts of sulfuric acid esters, $RO-SO_3Na$, or true alkylsulfonates, $R-SO_3Na$

   1/L Products obtained by sulfonation of secondary alcohols

   1/M Products obtained by mineral esterification agents other than sulfuric acid (e.g., phosphorus esters)

   1/N Amino carboxylic acids containing more than one —COOH group

   1/O Sulfonated aromatic hydrocarbons

   1/P Sulfonated derivatives of nonionic surfactants

1/Q Products obtained by starting with paraffin hydrocarbons

1/R Products based on waste lye (sulfite liquor) and cellulose

1/S Other anionic active substances not classified

In 1960, Rosen and Goldsmith [5] published their book on the analysis of surfactants, in which they reviewed the published classification schemes and found that none of them were suitable for the analyst. They therefore devised a scheme for analytical purposes. This was based on a primary breakdown according to the charge type and the elements present, with further subdivisions according to functional groups. The classification remained unchanged in the second edition of their book [6].

The main anionic groups listed are:

I.B.   Anionics containing metal, no N, S, or P

II.B.   Anionics containing S and metal, no N or P

III.B.   Anionics and ampholytics containing N and metal, no S or P

IV.A.1.   Anionics containing S, no N or P in the anion; N in the cation, no metal

IV.A.2.   Anionics containing N and S, no P in the anion; N in the cation, no metal

IV.B.   Anionics and ampholytics containing N, S, and metal, no P

V.B.   Anionics containing P, no N or S

V.C.1.   Anionics containing P, no N or S in the anion; N in the cation

V.C.2.   Anionics, cationics and ampholytics containing N and P, no S in the surface-active ion

Hummel [7] used a letter-number system for the characterization and identification of surfactants with infrared spectroscopy, the one system serving a dual purpose as illustrated by the following example for sulfosuccinic acid esters:

Classification

An   Anion-active

III   Sulfonate

A   Aliphatic

5   Several carboxylic groups in the molecule

b   Carboxylic groups esterified

Identification

An	Precipitates with cationic reagent
III	Intense IR absorption at 8.2 and 8.5 μm and at about 9.5 μm; qualitative test for sulfur
A	Absence of aromatic absorption
5b	Intense absorption at about 5.8 μm; after saponification, absorption of ionized carboxyl group at about 6.4 μm.

One of the results of the 1st World Conference of Detergency and Surfactant Products held in Paris in 1959 was the formation of the Comité International de la Détergence (CID). In 1960, at the 3rd International Congress of Surface Activity [8] this committee proposed an advanced and comprehensive scheme based on coding the individual groups and their position in the molecule into a 4-group 12-column system. The nomenclature [9] is compatible with Dyson's notation and enumeration system [10] which is the basis of the IUPAC recommended international chemical notation [11]. The International Organization for Standardization (ISO) started work to convert this into an international standard, publishing draft recommendations in 1961 [12]. Recently the aim to convert this system into an international standard has been shelved on the understanding that steps be taken to ascertain whether there is a demand in technical circles for a standard system for the classification of surfactants.

The scheme is complex and details of the latest version are given by Langmann and Hofmann [13]. The four groups describe the main and complementary characteristics of the hydrophilic and hydrophobic parts of the molecule and each group consists of three figures. Each column of figures can have up to ten options, including zero if that particular characteristic is not present in the molecule. For example, a sodium soap is represented as follows: 100 000 000 100. The first figure in the first group represents the hydrophilic, anionic carboxylate group. It is not cationic or nonionic, so the second and third figures in this group are zero. It has no hydrophobic surfactant group and this is represented by three zeros in the second triplet. Similarly, there are no complementary hydrophobic characteristics which accounts for the three zeros in the third group. The first figure in the fourth triplet represents the alkali metal salt (the complementary hydrophilic group). An ammonium salt would have been assigned as 7 in this column. The coding for monoethanolamine alkyl sulfate is 200 000 000 800 and that for sodium dialkylsulfosuccinate is 300 161 291 100, indicative of the more complex structure and the presence of complementary groups. This type of system is suitable for transfer of information to punched cards.

At the 4th International Congress of Surface Activity [9] a simplified, and therefore more restricted, version was proposed and further discussed at the 5th International Congress [14]. This simplified scheme is a five-column system consisting of one letter and four numerals.

Column 1.   The letter A (anionic), C (cationic), or N (nonionic) for
            ionic character

Column 2.   First figure for the constitution of the hydrophilic group

Column 3.   Second figure for the constitution of the hydrophobic group

Column 4.   Third figure for an intermediate functional group.

Column 5.   Fourth figure for the complementary character of the
            hydrophilic part.

A few examples will illustrate the application of the general and simplified schemes.

	General	Simplified
Sodium soaps	100 000 000 100	A1001
Monoethanolamine alkylsulfates	200 000 000 800	A2008
Sodium dialkylsulfosuccinates	300 161 291 100	A3021

For the formulator, a more practical classification would be a scheme based on commercially available products. This type of system is only of any real value if it is international and kept up to date. The McCutcheon Index [15] fulfills these requirements since it is published annually and recent issues have contained an alphabetical product guide which is updated according to commercial trends. It is now issued in two parts, the North American (which includes the classifications) and the International. The anionic surfactant groups included in the commercial product classification are listed alphabetically as follows:

Alcohols, ethoxylated monohydric sulfated

Alcohols, ethoxylated polyhydric sulfated

Alcohols, sulfates

Alkanesulfonates

Alkylsulfonates

Alkylarylsulfonates

Amines and amides, sulfonates

Arylsulfonates

Cyclic ethers, alkylphenols, ethoxylated sulfates, and sulfonates

Diphenylsulfonate derivatives

Fatty acids, soaps

Fatty acid esters, sulfonated

Naphthalene- and alkylnaphthalenesulfonates

Condensed naphthalenes, sulfonates

Oils and fatty acids, sulfates and sulfonates

Olefin sulfates and sulfonates

Oligomeric, anionic

Petroleum sulfonates

Phosphates, alcohol ethoxylates

Phosphates, alkyl acid

Phosphates, esters

Phosphates, ether ethoxylates

Succinates, sulfo derivatives

Taurates, also amide sulfates

The McCutcheon Index [15] contains a second classification system, also of particular value to the formulator, in which all surfactants are ranked according to their HLB value (hydrophilic lypophilic balance). To summarize, the classification schemes available and their main purpose are as follows:

Reference	Type of system	Application
CID (ISO) [13]	4-group 12-column	Comprehensive (punched cards)
CID [14]	One-letter 4-figure	Simplified
McCutcheon [15]	List of commercial product types	Formulation
McCutcheon [15]	Ranking of HLB values	Formulation
Hummel [7]	Letter-number system	Identification by IR spectroscopy
Rosen and Goldsmith [6]	Letter-number system	Analysis

II. DETECTION

Phenomena such as foaming, lowering of surface tension, etc., are indicative of the presence of surface-active agents in general but are not specific for anionic surfactants. However, anionic surfactants have a negative ionic charge and there are numerous detection methods which take advantage of this property and are hence specific for them.

The methods discussed here are, for the most part, suitable for detection purposes without prior separation from other surfactant species. There will, of course, be exceptions and results may be inconclusive. Under these circumstances, it is preferable to follow one of the separation methods and retest for the presence of anionic surfactant in the appropriate fraction. Once separation is achieved, it is possible to use general as well as specific methods because other types of surfactant will have been removed. Confirmatory tests are also advisable.

By far the greater number of methods reported in the literature involve the use of dyes but there are others based on precipitation, polarography, etc., which are included here.

Many of the detection methods for anionic surfactants can also form the basis of quantitative estimations, and these are described in Chapters 5 and 6.

The extensive use of anionic detergents has meant that there is considerable interest in their analysis and many workers have reported their results in the literature. Many of these are repetitive and for the most part reference is made here to the earliest work and to the more significant modifications and applications. There have also been many review articles and only selected references to the more comprehensive reviews are cited here.

A. Antagonistic Reactions

Anionic surfactants contain a negatively charged amphipathic ion which can react with compounds possessing a positively charged hydrophobic ion. Tschoegl [16] has called this an "antagonistic" reaction.

$$R^-M^+ + R'^+X^- = R^-R'^+ + M^+X^-$$

where

R^- = negatively charged amphipathic ion

R'^+ = positively charged hydrophobic ion

M^+ = H^+, Na^+, K^+, NH_4^+, etc.

X^- = monovalent acid ion, e.g., Cl^-, Br^-, CH_3COO^-, etc.

R'^+X^- can be either a basic dye or a cationic surfactant or a suitable salt of an inorganic acid. The product $R^-R'^+$ is usually water insoluble but often solvent soluble. This reaction and its variations form the basis of many of the detection methods.

1. Detection by Precipitation

Since the product of the antagonistic reaction is usually water insoluble this property can be utilized for a detection method.

a. Precipitation of Basic Dye Complex. Jones [17] first reported that colored, water-insoluble salts were formed between methylene blue and anionic surfactants and that the reaction product was soluble in organic solvents such as chloroform. Edwards et al. [18] filtered off the precipitated complex formed between sodium cetyl sulfate in aqueous solution and methylene blue solution acidified with sulfuric acid, then dissolved the precipitate in 70% aqueous ethanol solution for subsequent colorimetric determination.

The precipitation of anionic dye complexes is dependent on the concentration of the surfactant. In general, when the concentration is below the critical micelle concentration (CMC) precipitation occurs but at, or above, the CMC the complex is dispersed. Meguro [19] noted that the addition of a solution of malachite green oxalate to a very dilute solution of sodium dodecyl sulfate resulted in coagulation, but with a more concentrated detergent solution dispersion occurred. Later Kondo et al. [20] studied the effects of concentration and pH on the flocculation and deflocculation of basic dyes with sodium dodecyl sulfate. Colored flocculates that are soluble in chloroform and ethyl acetate can be formed with acridine yellow, methylene blue, and congo red, and defloculation occurs in the presence of excess sodium dodecyl sulfate. Methyl violet and methyl green are not flocculated even though an anionic dye complex, extractable with organic solvent, is formed, attributed to resonance in the complex. Rhodamine G in the presence of dodecyl sulfate becomes fluorescent and flocculation and deflocculation occur at specific dye concentrations.

Burger [21] used the precipitation reaction to detect very small amounts of anionic surfactants in aqueous solution. His test procedure is as follows: Shake the aqueous test solution, containing up to 20 ppm surfactant, with an aqueous indicator solution containing 0.05% of methylene blue and petroleum ether containing 1% cetyl alcohol. Allow the mixture to separate. Blue particles at the interface of the two-phase system indicate the presence of anionic surfactants. Burger did in fact use a mixed indicator of methylene blue and pyrocatechol violet which allowed differentiation between anionic and cationic surfactants. Blue particles from the basic dye (methylene blue) and anionic surfactant indicated the presence of anionics, and yellow particles from the acid dye (pyrocatechol violet) and cationic surfactant indicated the presence of cationics. There is no mention of interfering substances but the presence of nonionic surfactants gives rise to an emulsion layer at the interface.

b. Precipitation with Cationic Surfactants. Precipitation of one ionic surfactant by another with an opposing charge in aqueous solution has been known for a long time. Most anionic/cationic surfactant complexes are water insoluble, and this property can be utilized for detection purposes. Caution is necessary for this type of test because the anionic/cationic complex is soluble in excess of anionic or cationic surfactant.

General Test

Reagent: 1% aqueous solution of cationic surfactant
Procedure: Add the 1% cationic solution dropwise to a 1% aqueous solution of the sample. Observe whether a precipitate forms which disperses on shaking. Continue dropwise addition until the precipitate remains. The test sample should not contain inorganic salts. These are best eliminated by chloroform extraction, using a 1% aqueous solution of the dried chloroform extract for the test.

Wickbold [22] utilized this test for determination of alkylarylsulfonates and sulfosuccinate esters. He noted that, just before clarification of the solution indicating solubilization in the presence of excess cationic titrant, these two types of surfactant exhibited a turbidity maximum which could be detected photometrically using light-dispersion measurements. Other anionics such as alkyl sulfates did not show a turbidity maximum and could not be detected or determined by this method. Bennewitz and Fiedler [23] converted this into a more general method by taking the sudden clearing of the turbid solution as the end point. Although not essential, they included an indicator (bromophenol blue) which made the visual observation of the end point easier. In their method, the cationic titrant is added while the mixture is stirred, turbidity increases, and the color changes from violet to greenish to white and finally to blue. At the white stage it is advisable to stop the stirring for a short while after each addition in order to detect the sudden appearance of blue colored flakes or drops which rapidly precipitate. This type of method serves as detection and estimation at the same time.

c. Precipitation with Antagonistic Electrolytes. The third class of compounds which can react in an antagonistic manner with anionic surfactants and are listed by Tschoegl [16], are the water-insoluble amine salts of inorganic acids. Various amine salts have been used for quantitative estimation as well as for detection.

Kling and Pueschel [24] used benzidine hydrochloride; Marron and Schifferli [25] used p-toluidine hydrochloride. Another alternative is phenylhydrazine hydrochloride. The precipitate is usually soluble in organic solvents such as chloroform, carbon tetrachloride, ethanol, etc.

Other electrolytes also precipitate anionic detergents, for example aluminum salts. Wurzschmitt [26] used aluminum acetate. Rosen and Goldsmith (Ref. 6, p. 30) prefer Blankorol B (BASF Ludwigshafen), described by the manufacturers as a "powdered water-soluble basic alumina," because it precipitates anionics even from an anionic/cationic complex solubilized by excess cationic surfactant.

Linari [27] precipitated sodium lauryl sulfate from solution by the addition of 25% aqueous potassium nitrate solution. Takayama [28] used barium chloride solution to precipitate the sparingly soluble barium salts of anionic surfactants from anionic/nonionic mixtures. The barium salts could be identified subsequently by infrared spectroscopy.

2. Detection by Solvent Extraction of Anionic
    Surfactant/Dye Complexes

As mentioned earlier, the product of the antagonistic reaction is usually soluble in organic solvents and if a basic dye is used, the anionic surfactant/dye complex can be extracted into the organic solvent to give a visual detection method.

Jones [17] reported that the precipitate obtained by reacting anionic surface-active compounds with methylene blue was soluble in chloroform and that this property could be used for qualitative and quantitative purposes.

General Test

   Reagents: Methylene blue, 0.25% solution in ethanol; hydrochloric acid; chloroform.
   Procedure: To about 20 ml of test solution add a slight excess of hydrochloric acid. Add 1 ml of methylene blue solution and 20 ml chloroform. Shake well and allow the layers to separate. A deep blue color in the chloroform (lower) layer indicates the presence of anionic surfactant.

Soaps give little or no color because their surfactant properties are destroyed by the addition of acid. Methylene blue has probably been more widely used than any other dyestuff over the years but other basic dyes have also been used. Karush and Sonenberg [29] complexed long-chain alkyl sulfates with rosaniline or para-rosaniline hydrochloride, extracting the complex into a 1:1 mixture of ethyl acetate and chloroform. The dye was prepared in a pH 6.1 buffer and the sensitivity was found to increase with increasing chain length of the alkyl group, which they attributed to the greater solubility of the larger molecules in the organic solvent. Wallin [30] used basic fuchsin which is composed of rosaniline and pararosaniline. He carried out the reaction in acid solution and although he did not definitely establish the fact, he believed that only one component of basic fuchsin took part in the reaction. Wallin preferred basic fuchsin, even though it is less specific than methylene blue, because of its insolubility in chloroform, which enabled him to use chloroform as a blank in his quantitative method. It is suspected that commercial supplies of methylene blue available at that

time contained impurities which were chloroform soluble. Pure methylene
blue in acid solution is not extractable by chloroform.

For soaps and alkyl sulfates, Salton and Alexander [31] used pinacyanol
bromide, and Klevens [32] used pinacyanol chloride and anisoline. Yamagishi and Yanagisawa [33] reacted sulfated and sulfonated anionics with
Pinacryptol yellow in acid solution (below pH 3) extracting the complex into
chloroform. However, high concentrations of inorganic halides, potassium
cyanide, and proteins caused interference. Milwidsky and Holtzman [34]
and later Bares [35] used bromocresol green for their quantitative determination of soaps and anionic surfactant mixtures. Saito et al. [36] used
Rhodamine 6G, extracting into benzene. The use of Azure A and extraction
into chloroform was proposed by van Steveninck and Riemersma [37]. This
has advantages over methylene blue, pinacyanol salts, and basic fuchsin
because it is much less sensitive to interference from other anions. They
also noted that with a mixture of C_{12} and C_{17} alkyl sulfates, monochloro-
benzene preferentially extracted the C_{17}/Azure A complex, whereas chloroform extracted both, and suggested the possibility of analyzing mixtures of
alkyl sulfates with differing chain lengths by this method.

McGuire et al. [38] used Toluidine Blue O for a field test to detect anionic surfactants in well waters.

General Test

Reagent: Prepare dye solution, by mixing 200 mg sodium arsenite,
40 g sodium dihydrogen phosphate monohydrate, and 20 mg Toluidine
Blue O. Dissolve the mixture in 200 ml of 1 N sulfuric acid. Extract
three times with 20 ml of chloroform to remove impurities.
Procedure: Add 0.4 ml of dye solution to a 10-ml sample of well water.
Add 2.5 ml chloroform, shake in a stoppered tube for 30 sec; if the
chloroform layer is colored blue, anionic surfactant is present. As
little as 0.05 mg of alkylbenzenesulfonate per liter can be detected and
measured.

Most of the preceding detection methods have been carried out in acid
solutions, which has the advantage that soaps are protonated into the corresponding fatty acid thus losing their surface activity and permitting ex-
clusive detection of anionic surfactants that are stable in cold acid. It is
possible to carry out detection methods in alkaline solution in which case
fatty acid soaps also react. Turney and Cannell [39] developed a method
based on the oxidation of methylene blue chloride to dimethylthionoline (a
red dye) in the presence of chloroform and sodium hydroxide. Geyer [40]
modified this method using an alkaline phosphate buffer instead of sodium
hydroxide. Uno and Miyajima [41] used neutral red at a pH of 8.5 as well
as at pH 5 [42].

3. Detection by Dye Transfer

a. With Basic Dyes. In a two-phase system of aqueous and organic solvent layers it is possible to equalize the color in each layer using a mixture of dye and surfactant/dye complex. Any change, for example, by addition of more surfactant, will result in transfer of color from one layer to the other. Weatherburn [43] mixed together an aqueous solution of methylene blue and a chloroform solution of anionic surfactant/methylene blue complex, adjusting the amounts to give equal color depth in both phases. The addition of an anionic detergent means that more anionic/methylene blue complex is formed, which in turn is chloroform soluble; hence the chloroform layer becomes darker and the aqueous phase paler. Conversely, the addition of a cationic surfactant gives rise to a paler chloroform layer and a darker aqueous layer. Most of the basic dyes mentioned in the preceding section dealing with detection by solvent extraction of anionic/dye complexes can be used for the dye transfer test.

General Test

Reagents: Prepare dye solution with 0.03 g methylene blue, 12 g concentrated sulfuric acid, and 50 g anhydrous sodium sulfate, diluted to 1 liter with distilled water; anionic surfactant solution with 0.05% solution of dioctylsulfosuccinate in distilled water; and chloroform.
Procedure: Place 8 ml of dye solution and 5 ml chloroform in a 25 ml glass-stoppered cylinder or test tube. Add the anionic surfactant drop by drop, stopper the tube, shake and allow to separate, until the two phases are equal in color when viewed in reflected light (about 10-12 drops required). Add 2 ml of a 0.1% solution of the test substance, shake, and allow the layers to separate. If the chloroform layer is deeper in color and the aqueous phase paler, the unknown is anionic. Soaps are not detected because the reagent is acidic and phosphates or silicates do not interfere. Cationic surfactants will give a deeper colored aqueous phase, whereas nonionic surfactants will cause emulsification but not affect the color of either phase.

b. With Acid Dyes. Acid dyes can also be used for this type of test, if the color of the two layers can be equalized by the addition of a cationic surfactant. The presence of anionic surfactant is detected when the aqueous phase becomes darker and the chloroform layer paler.

A number of workers have used acid dyes in two-phase systems to detect the end point for quantitative determination. For example:

Bromophenol blue—Barr, Oliver, and Stubbings [44]

Brilliant Blue FCF—Lewis and Herndon [45]

Azophloxine—Edwards and Ginn [46]

Metanil yellow—Bennewitz [47]

Erythrosine and Eosine—Aoki and Iwayama [48]

Kristall ponceaux 6R extra—Stojkovic and Kukovec [49].

    c. With Dye Mixtures. The dye transfer method can be taken a stage further by using a carefully selected mixture of different colored acid and basic dyestuffs to form a 1:1 acid/base dye complex. If the unknown test sample in aqueous solution is mixed with a solution of the acid/base dye complex in the presence of chloroform and shaken, appearance of the basic dye color in the organic phase indicates the presence of anionic surfactant. If the unknown sample contains cationic surfactant the color of the acid dye will appear in the organic layer. Aoki and Iwayama [50] reported that a salt is formed between a sulfophthalein dye and neutral red. Examples of suitable sulfophthalein dyes were bromophenol blue and bromocresol green. The salt is a 1:1 molar complex and is chloroform soluble. When buffered at pH 6 to 7 the color is pale yellow. In the presence of an anionic surfactant the chloroform layer becomes red. A green color indicates the presence of cationic surfactant.

    The most successful and widely used mixture is that first introduced by Holness and Stone [51] which consists of an acid dye, Disulphine Blue V200, and a pink-colored quaternary ammonium compound, dimidium bromide (2,7-diamino-10-methyl-9-phenylphenanthridinium bromide), which acts as the basic dye. They used the dye mixture at two different pH values, 1.99 and 8.6. Neither of the dyestuffs at either pH are extractable into chloroform in the absence of surfactants. Using the pH 1.99 indicator, the chloroform extract was pink in the presence of sulfonated and sulfated surfactants and colorless in the presence of soaps. At pH 8.6, fatty acid soaps as well as sulfonated and sulfated anionics gave a pink chloroform extract.

4. Detection by Color Change of the Dye Complex

The change of color of pH indicators by anionic surfactants without change of pH was reported by Corrin and Harkins [52] and Klevens [32], who noted that this effect occurs at, above, or near the CMC and can involve change in color or fluorescence or both. The color without surfactant, or with surfactant concentration below the CMC, is called the "molecular color," and the color resulting from surfactant concentrations above the CMC is called the "micellar color." Pinacyanol bromide [31] and pinacyanol chloride [32, 53] have a pink molecular and a blue micellar color. Thymol blue [54] gives a definite red-violet micellar color with anionic surfactants. The reaction is very distinct and forms a useful general detection test.

General Test

    Reagents: Thymol blue solution 1%, hydrochloric acid 0.005 M

Procedure: Add 5 ml of neutral aqueous test solution to 5 ml of 0.005 M hydrochloric acid containing 3 drops of thymol blue solution. A change in color to red-violet indicates the presence of anionic surfactant.

Bennewitz [47] examined the color changes which occur with a number of dyes and different types of surfactant, tabulating the results. Thymol blue, cresol red, Tropeolin 00, metanil yellow, bromophenol blue, bromo-cresol green, and bromothymol blue all showed color changes in the presence of anionic surfactants; his preferred indicator was thymol blue.

Aoki and Iwayama [48] used Toluidine blue at pH 8 in a chloroform/aqueous two-phase system. The molecular color in the chloroform layer is crimson but in the presence of anionic surfactant the blue micellar color appears. They also used erythrosine and eosine at pH 4.5 and noted fluores-cence changes. Using sodium eosinate as a fluorescent indicator, Dolezil [55] noted that there is an increase of fluorescence at the CMC when the examination is carried out under UV light. Rhodamine 6G [36] becomes fluorescent in the presence of dodecyl sulfate and other anionic surfactants.

More recently, Egginger [56] has reported color differences with dif-ferent classes of surfactant. To a 1% neutral test solution he adds 3 ml of cresol red solution (3 drops of 1% cresol red solution and 10 ml 0.1 N hydro-chloric acid). In the presence of anionic surfactants the solution is red, whereas cationics give a yellow color and nonionics an orange color.

B. Detection by Color Restoration

Abramovich [57] decolorized Brilliant Green or methyl violet with sodium sulfite and found that in the presence of anionic surfactants the original color was restored.

General Test

Reagent: Mix 1 vol of aqueous dye solution (Brilliant Green or methyl violet) with 2 vol of fresh 5% sodium sulfite (Na_2SO_3) solution.
Procedure: To 5 ml of test solution add 5 ml of reagent. Shake well to mix. Restoration of dye color indicates the presence of surfactant. This test is very sensitive so all glassware and reagents must be free from surfactants and oxidizing substances.

C. Detection by Addition of Reducing Agent

Renault and Bigot [58], appreciating the interferences which affect the pre-ceding method, proposed a method using a colorless cationic compound which can be converted to a colored compound by the addition of a reducing agent. The reaction is characteristic for long-chain sulfur-containing anionic surfactants. Proteins are removed by precipitation with 10% thiosalicylic acid, persalts are removed by boiling with alkali, and the only interference is from high levels of thiocyanates.

General Test

Reagents: pH 2.5 buffer, 7.5 g glycine, 5.8 g sodium chloride dissolved in 1 liter of dilute hydrochloric acid; triphenyltetrazolium (TPT) chloride, 1% solution in water, freshly prepared; sodium hydrosulfite, 1% solution in pH 7 phosphate buffer; cetyltrimethylammonium bromide, 0.5% solution in water; dibromoethane.

Procedure: To 40 ml of neutral test solution add 1 ml of pH 2.5 buffer and 1 ml TPT chloride solution. Extract twice with 5 ml dibromoethane. Wash the organic layer 3 times with 10 ml of the buffer diluted to 30%. Mix the organic layer with 2 ml of cetyltrimethylammonium bromide solution. Add 3 drops of sodium hydrosulfite solution. The development of a pink or red color indicates the presence of anionic surfactant. The test is suitable for the detection of anionic surfactants in wines, soda water, beer, and milk.

D. Detection by Chelatometry

Chelatometry has been used mainly for quantitative determination but it can also be used as a detection method. Taylor and Fryer [59] report that complexes are formed between iron(II) bipyridyl and anionic surfactants, according to the following equation:

$$\left[Fe(2,2'\text{-bipyridyl})_3\right]^{2+} + 2\ (anionic)^- \longrightarrow \left[Fe(2,2'\text{-bipyridyl})_3\right] (anionic)_2$$

The complex is completely soluble in chloroform. Ferroin reacts in a similar manner and has a higher molar extinction coefficient. Since both complexes are red in color they can be used for detection purposes as well as for quantitative estimation.

Courtot-Coupez and Le Bihan [60] formed an ion pair between o-phenanthroline Cu(II) cation and an anionic surfactant, extracting the complex into methyl isobutyl ketone and determining the copper present by atomic absorption spectroscopy. This method could be used for detection.

Fowler and Steele [61] determined anionic surfactants, such as C_{14}-C_{22}-fatty acid soaps, sodium C_{12}-C_{14}-alkyl sulfates, sulfonates, and dinonyl phosphoric acid, by treatment with ethanolamine-buffered copper triethylenetetramine reagent and extraction into a mixture of cyclohexane and isobutanol. The organic extract was treated with diethylammonium diethyldithiocarbamate, followed by spectrophotometric determination of the copper complex at 435 nm. This method could be adapted for detection purposes.

E. Detection by Polarography

There are a number of polarographic methods for the determination of surfactants but although these are often very sensitive they are not particularly selective. Properties such as suppression of oxygen maxima, shifting of oxygen maxima to higher potentials, and slowing down of the formation and descent of mercury droplets have all been used for quantification. A method that could possibly be used for detection has been devised by Buchanan and Griffith [62], based on the reduction in height of the polarographic step of methylene blue, buffered at pH 4.5, by the addition of anionic surfactant.

F. Distinguishing Tests

1. Distinction Between Soaps and Anionic Surfactants

Soaps are anionic in character but in acid conditions they change to the corresponding fatty acid and lose their surface activity. All anionic surfactants retain their surface activity in cold acidic solutions at least long enough for testing purposes. In general, any of the detection methods mentioned which are carried out at alkaline pH will detect soaps, but all those in acid solution will not.

2. Distinction Between Anionic and Cationic Surfactants

Most of the detection methods listed involve the ionic character of the surfactant, thus there is no confusion with nonionic surfactants. Some of the methods are suitable for distinguishing between anionic and cationic surfactants and this has been mentioned under the appropriate headings.
The main methods are:

Precipitation, Burger [21] (Sec. II.A.1.a)

Dye transfer, Weatherburn [43] (Sec. II.A.3.a)

Mixed dye transfer, Aoki and Iwayama [50] (Sec. II.A.3.c)
                    Holness and Stone [51] (Sec. II.A.3.c)

Color differences, Egginger [56] (Sec. II.A.4)

3. Distinction Between Sulfated and Sulfonated
   Anionic Surfactants

This topic will be discussed in detail in other chapters but it is worth noting
here the part of the general scheme by Greenberg [63] that differentiates
between alkyl sulfates and sulfonates.

General Test

> Reagents: Methyl yellow solution—mix 25 g citric acid and 0.2 g
> methyl yellow with 25 ml water; bring to the boil and filter. Benzidine
> solution—mix 10 g citric acid and 0.5 g benzidine with 10 ml water and
> bring to the boil; separately dissolve 15 g tartaric acid in 15 ml boiling
> water; combine the two solutions, boil, cool, and filter (if necessary).
> Tartaric acid and sodium metavanadate (2% aqueous solutions).
> Procedure: To 10 ml of test solution add 2 drops of methyl yellow
> solution. Acidify if necessary with tartaric acid. A cherry red color
> indicates anionic surfactants, i.e., sulfates and sulfonates (orange
> color indicates cationics or nonionics). To another 10 ml of test solu-
> tion add 2 drops of benzidine solution. Acidify if necessary. Add 2
> drops of sodium metavanadate solution and mix. A pale yellow color
> indicates sulfated anionics, a permanganate-purple color after 3 min
> indicates sulfonated anionics (dark green color—nonionics; opaque
> yellow—cationics). Add 2 drops of methyl yellow solution to the mix-
> ture. An unchanged purple indicates sulfonated anionics, cherry red in-
> dicates sulfated anionics (dark red-brown—nonionics; orange—cationics).

This type of test is not suitable for mixtures of anionic surfactants.

III. SEPARATION AND ISOLATION

After detection methods have shown the presence of anionic surfactant, the
next step is the identification or determination of the materials present.
Generally this requires an intermediate separation step to remove possible
interfering substances because of the often complex nature of the sample in
which the anionic may occur. The separation step also has the advantage of
concentrating the surfactant before further identification techniques are used.

A. Occurrence of Anionic Surfactants

Anionic surfactants can be used in a wide variety of products, including industrial and domestic detergents, emulsions, foods, cosmetics and toiletries, dyestuffs, paints, oils, polishes, and inks and may finally be dis-
charged in effluent, waste water, and sewage. However, by adopting the

simple classification of Rosen and Goldsmith [6], a systematic approach to
the isolation of anionics may be taken. Accordingly, surfactants may occur
in media consisting of:

Water or aqueous solutions

Volatile solvents or their aqueous emulsions

Nonvolatile organic matter in solutions or emulsions

Nonvolatile inorganic matter

Nonvolatile organic matter

Insolubles (pigments, resins, etc.)

The analytical procedures most widely used fall into two categories,
namely, extraction methods, by solvents or foaming; and chromatographic
methods, including ion exchange, paper, thin layer, and column.

B. Extraction Methods

1. Solvent Extraction

   a. General Considerations. Solvent extraction has been widely used
for the separation of anionic surfactants since it is simple to perform and
requires a minimum of equipment. However, no one solvent appears suitable for the extraction of all types of anionic surfactant, although they tend
to have a greater solubility in the more polar solvents, such as water,
alcohols, ketones, esters, ethers, and chlorinated hydrocarbons. Some
useful general factors affecting solubility are given by Rosen and Goldsmith [6].

   1. Length of the hydrophobic chain. In a homologous series, water
       solubility decreases and nonpolar solvent solubility increases with
       increasing chain length.

   2. Branching and presence of aromatic rings have only a minor effect
       on water solubility.

   3. Bridging functions such as ether oxygen, and ester, amide, and
       sulfone groups tend to make the surfactant more polar but also
       extend the solubility range.

   4. Number and type of hydrophilic groups. Increasing the number of
       polar groups increases water solubility and may reduce solubility
       in nonpolar solvents. Water solubility increases by progression
       from carboxyl to sulfonic or sulfuric ester substitution.

5. Nature of the cation. Sodium soaps are less soluble in water than
   potassium soaps, while the reverse holds for the salts of sulfonates.
   Amine cations increase the solubility in organic solvents.

6. Length of the oxyethylene chain. Water solubility increases with
   increasing length of the polyoxyethylene chain whereas solubility
   in aliphatic hydrocarbon solvents is decreased.

7. Effect of pH. Water solubility is lowered if free acids are formed
   because they are more soluble in solvents than water.

8. Electrolytes. These generally lower water solubility and tend to
   displace surfactants into immiscible organic solvents.

9. Presence of water. Small quantities of water will greatly increase
   the solubility of surfactants in polar solvents. Simultaneously the
   surfactants increase the water miscibility with solvents. Of
   greater importance is the fact that alcohols and acetone may extract significant amounts of inorganic materials (especially NaOH,
   KOH, and NaCl) if water is present.

10. Mixtures of surfactants. The solubility of surfactants in water or
    solvents is often modified by the presence of a second surfactant.
    Thus cationics produce solvent-soluble complexes with anionics.
    Alkanolamides solubilize sulfonates and sulfates in ether and soaps
    solubilize monoglycerides in water.

   b. Extraction from Aqueous Solutions. Although anionic surfactants
can be isolated from aqueous solutions or emulsions by evaporation, if the
concentration of anionic is low a large volume of sample may be required.
This can be both inconvenient and tedious. Extraction with an immiscible
solvent then offers a more practical alternative.
   Miller et al. [64] describe the use of n-butanol for the quantitative extraction of several types of anionic surfactant (sodium lauryl sulfate, sodium
sec-alkyl sulfate, sulfated fatty acid monoglyceride and sodium alkylarylsulfonate) from their dilute (0.4-4%) aqueous solutions. They increased the
efficiency of extraction by the addition of sodium carbonate (4% by weight of
the aqueous sample) to salt out the surfactant and decrease its water solubility. The carbonate also served to reduce hydrolysis of the alkyl sulfates.
Evaporation of the extract was made in vacuo and the residue was found to
contain less than 0.05% of the added sodium carbonate.
   Izawa and Kimura [65] found that extraction of primary and secondary
alkyl sulfates and alkylbenzenesulfonates by butanol gave values similar to
those obtained by the Weatherburn [43] two-phase titration method. The
Epton method [66] was found to give lower values.
   Shang [67] described the use of several water-immiscible organic
liquids (solubility less than 0.1%) for the extraction of alkylbenzenesulfonates from water including cyclohexane, kerosine, naphtha, benzene, toluene,

benzophenone, and benzyl chloride, although hydrocarbon types were preferred. After the extraction the organic layer was treated with lime to
precipitate the alkylbenzenesulfonate (ABS) as the calcium salt, which was
filtered off.

Very low levels of alkylbenzenesulfonates in waters and sewage have
been determined by extraction as solvent-soluble complexes. Sallee et al.
[68] and Fairing and Short [69] converted the ABS to its 1-methylheptylamine
salt by adjusting the pH of the sample to 7.5 and adding a 0.1% solution of
methylheptylamine in chloroform. The ABS/amine complex was extracted
into the chloroform phase which was then separated for determination of the
ABS present by infrared or visible spectrophotometry. Levels of less than
1 ppb in water and 50 ppb in sewage were detectable.

Ogden et al. [70] and Swisher [71] used a similar procedure but preferred the use of n-heptylamine, whereas Matsumoto et al. [72] used propyl-
amine. The complexes arising from these latter amines were extracted
with a low-boiling petroleum fraction.

Dye complexes are discussed in Sec. II.A.2; many of these are also
solvent extractable.

c. Extraction from Nonvolatile Inorganic Matter. One of the most
important applications for many anionic surfactants is in detergent powders
and liquids for household use and general cleaning purposes. Often the
"active" surfactant is supplemented by the addition of auxiliary surface-
active materials and inorganic builders. These substances are included to
improve the action of the product with regard to detergency, foaming, dirt
suspension, etc., and may include alkanolamides, sodium carboxymethylcellulose, condensed phosphates, silicates, perborates, carbonates, chlorides, and sulfates. Solvent extraction methods to isolate the active sur-
factant(s) have been commonly employed for their analysis.

Again the whole range of polar solvents has been used, although ethanol
perhaps most often. Generally, the inorganic salts are almost insoluble in
the anhydrous solvents but, as already mentioned, the presence of even
small amounts of water may increase the solubility of surface-active substances but the quantity needs to be controlled. This is achieved by extracting the thoroughly dried sample with 95% ethanol.

(1) Extraction with 95% Ethanol. Ethanol (95%) has been used in several different ways for extraction.

Simple digestion. The sample is extracted directly with hot alcohol
[73, 74] or is heated under reflux [75]. The insoluble matter is allowed to
settle and the alcohol is decanted through a sintered glass funnel. After
evaporation of the alcohol, the solids are reextracted several times more
with a smaller volume of 95% ethanol, then dried to remove the alcohol and
finally treated by precipitation/extraction. The combined alcohol extracts
may be evaporated to dryness on a steam bath, then dried to constant weight
either in an oven at 100°C or by adding dry acetone and reevaporating on the
steam bath. If the presence of sec-alkyl sulfates is suspected, the extract

should be neutralized to phenolphthalein before evaporation to prevent hydrolysis.

Soxhlet extraction. This method is similar to simple digestion except that the Soxhlet extractor helps to retain insoluble matter thus giving more efficient extraction [75-77]. The procedure is straightforward for powders, but liquids need special treatment. One method [77] is to evaporate the liquid on a steam bath, powder the dried residue, and then treat this like powder samples. If the sample will not dry to a powdery mass, the liquid can be mixed with fine sand or silica prior to drying on a steam bath.

Precipitation/extraction. In this method the low solubility of inorganic salts in alcohol is used [73, 77]; it is especially useful where the active ingredient is closely mixed with the inorganic matter (e.g., spray-dried products). In one procedure [73], the sample is dissolved in the minimum quantity of hot distilled water and the inorganic salts precipitated by slowly adding 95% ethanol with vigorous stirring. The solution is then heated to boiling on the steam bath and filtered, the insoluble material being washed several times with additional 95% ethanol.

In the second procedure [77], the sample is dissolved in a large volume of alcohol, filtered, and evaporated to dryness. The dried residue is then redissolved in absolute alcohol, filtered again, and the filtrate finally evaporated to constant weight.

Ross and Blank [78] suggest a modification of the extraction methods to remove sodium carbonate and borax which may also be extracted. After the alcohol solubles have been evaporated to dryness, they are reextracted with a 1:1 mixture of acetone and diethyl ether. The extract is then filtered, evaporated, and dried to constant weight at $80 \pm 2°C$.

(2) Extraction with Other Alcohols. Among the other solvents found suitable for extraction, absolute alcohol has been used by Balthazar [2] after thorough drying of the sample. Soxhlet extraction was used for powder samples and simple digestion for liquids. Wurzschmitt [79] preferred the use of absolute methanol, also for well-dried samples.

Vizern and Guillot [80] also used absolute ethanol for the extraction of sulfonated fatty acid amides. The sample (2 g) was dried at 85-90°C, taken up in portions of 50 ml of hot absolute ethanol, and then refluxed for 15 min. After the alcohol had been decanted the extraction was repeated twice more with separate 20-ml aliquots of alcohol.

Aqueous isopropanol (95%) was used by Weeks and Lewis [81], and also by Kortland and Dammers [82] in place of ethanol in the Berkowitz method [73] with good results. Weiss et al. [83] have used 50% aqueous isopropanol for the extraction of petroleum sulfonates and carboxylates from inorganic salts in a modification of the ASTM Method D855 [84]. After the sample has been dissolved in the aqueous alcohol, the solution is warmed to about 40-50°C and then saturated with sodium carbonate. On standing the isopropanol (containing the actives) separates and can be removed.

Butanol has been used by a number of workers including Gerber and Sporleder [85], Milwidsky [77], and Holness and Stone [51]. Holness and Stone originally used n-butanol following the method of Miller et al. [64] but found that the yield with this solvent was too small; tert-butanol gave superior results. Final traces of this alcohol were removed by adding n-hexane to the evaporated extract and reevaporating. Milwidsky describes the extraction of active ingredients with n-butanol, isobutanol, or sec-butanol:chloroform (1:1) in continuous extractors of his own design [86]. Because of the high boiling point of butanol he suggests the following method of evaporating the extract to prevent charring: The alcohol is evaporated to a volume of 2 to 3 ml on a hot plate; 2 ml of water are then added to form an azeotrope with the butanol boiling at about 90°C. Evaporation is continued until the sample is almost dry or until the odor of butanol has disappeared; the residue is finally dried in an oven at 105°C.

(3) Extraction with Other Solvents. In order to separate sodium salts of commercial alkylbenzenesulfonates from extraneous inorganic substances, Longman and Hilton [74] have used the fact that they are soluble in chloroform when dry. However, if toluene- or xylenesulfonates are also present, the method must be modified, for although these materials are insoluble in chloroform they remove a small amount of the alkylbenzenesulfonate and they also clog and slow down filtration. This difficulty was overcome by using dry acetone:absolute alcohol (1:1) in which the toluene- and xylenesulfonates are also extracted.

The alkylbenzenesulfonate can be further separated by extraction with diethyl ether from acid solution, taking advantage of the increased solubility of the free acid in the nonpolar solvent. This property has been applied by a number of workers for extracting sulfonates. Brooks et al. [87] used extraction with chloroform for petroleum sulfonates. House and Darragh [88] separated alkylarylsulfonates from inorganic salts by solution in 6 N hydrochloric acid and diethyl ether extraction, whereas Milwidsky [77] preferred 3 N hydrochloric acid. Simmons et al. [75] suggest the use of methyl isobutyl ketone which dissolves less water than ether.

Monoalkanolamine and trialkanolamine lauryl sulfates have been extracted from the amine salts by the use of hot dry acetone [74]. The extract, however, is contaminated with free alkanolamine, nondetergent organic matter and possibly lather improvers.

The ether solubility of the ammonium salts of sulfated and sulfonated anionics has been utilized by Hart [89]. He observed that the sodium or potassium salts of these surfactants were quantitatively converted to the ammonium salt when shaken with a concentrated solution of ammonium chloride or sulfate. The ammonium salt could then be isolated by extraction with ether.

Bergeron et al. [90] used ethanol, chloroform, or ethyl acetate for the extraction of anionics, whereas Shirolkar and Venkataraman [91] found

dioxane or ethyl acetate highly suitable. Jahn [92] was able to extract aliphatic alcoholsulfonates using a mixture of ethanol:trichloroethylene (1:1).

d. Extraction from Nonvolatile Organic Matter. As with extraction
from inorganic matter, there is no single extraction method suitable for all
applications within this group. It is therefore convenient to discuss the
methods that have been used in relation to the matrix involved.

(1) Extraction from Surfactant Raw Materials and By-products of
Manufacture. Methods of extraction of "actives" from nonsurfactant materials have been widely investigated because of their importance in the
analysis of commercial detergent products and a number of standard methods have been published; for example, ASTM [84], AOCS [93], Longman
and Hilton [74], and Simmons et al. [75]. The first step in these methods
is usually the separation of "actives" from material variously known as
"nondetergent organic matter," "fatty matter," or "free oil" and consisting
mainly of unsulfated, unsulfonated, unsaponified, or unreacted organic
matter. When only alkanolamine salts or small amounts of inorganic salts
are present in the sample, the separation can be performed by extracting a
50% aqueous ethanolic solution of the sample with hexane or light petroleum
ether with a boiling range of 40 to 60°C [73, 74, 94, 95]. Alternatively,
extractions from 20% isopropanol [95], 50% isopropanol [96], and 95%
ethanol [77] have been used. In commercial built powders, it is preferable
to first isolate total organic matter by extraction of the product with 90%
ethanol before petroleum ether extraction. The anionic surfactant remains
in the aqueous alcohol phase.

(2) Separation from Soap. When soap is present in the sample, it is
partly extracted into the petroleum ether phase, but can be separated by
washing it with 0.5 N aqueous alkali. If soap and anionic detergent are
present in admixture in the product, the soap may be separated by acidifying the aqueous ethanolic phase remaining after isolation of nondetergent
organic matter and extracting with diethyl ether [74]. Alternatively, the
soap may be extracted from a 50% aqueous ethanol solution of the sample
with petroleum ether [73, 77]. The soap fatty acids have also been isolated
by precipitation as their magnesium salts [97].

(3) Extraction from Product Additives. Some alkanolamides may also
be extracted into petroleum ether in which case the extract is washed with
70% aqueous ethanol to remove them. Milwidsky [77] notes that materials
of the stearyl alcohol sulfate type are also appreciably soluble in petroleum
ether.

Unsulfated material from products containing monoglyceride sulfate
or alkanolamide sulfate may not be soluble in petroleum ether. If alkanolamides are absent, this material may be extracted with diethyl ether from
a 30% aqueous ethanolic solution of the sample [74].

Kortland and Dammers [82] suggest removing alkanolamides, fatty
acid amides, and higher alcohols from anionics by repeated extraction of a
10% isopropanol solution of the sample with ether:hexane (1:1).

Separation of alkylbenzenesulfonates from toluene- or xylenesulfonates has been mentioned in Sec. III.B.1.c.(3). House and Darragh [88] found that multiple extractions of a sample with diethyl ether from 4 N hydrochloric acid solution quantitatively removed higher alkylbenzenesulfonates into the ether phase whereas lower-molecular-weight material, with less than four carbon atoms in the alkyl group, remained in the aqueous phase. The method was satisfactory for separation of benzene- and toluenesulfonates but was not as effective with xylenesulfonates.

(4) Extraction of Sulfonates as Sulfonic Acids. The extraction of sulfonates was also discussed in Sec. III.B.1.c.(3). Reutenauer [95] isolated, as the sulfonic acid, the "active" from sulfonated alkylaryls by two consecutive exhaustive extractions with benzene and amyl alcohol from hydrochloric acid solution. Milwidsky [77] used continuous extraction with diethyl ether from 3 N hydrochloric acid solution to extract monosulfonates but precautions were needed because of the appreciable solubility of hydrochloric acid in ether. Küpfer et al. [94] extracted alkanemonosulfonates with petroleum ether (30-60°C) from 50% aqueous ethanol acidified with hydrochloric acid.

In Milwidsky's method, the disulfonic acids formed are not soluble in ether, but are soluble in alcohols. Therefore, after removal of the monosulfonic acids the disulfonic acids could be extracted with n-butanol or isobutanol from the aqueous acid solution. The barium salts of the monosulfonic acids are precipitated by barium chloride solution from an aqueous solution of the sample [76]. The disulfonate salts are very soluble and remain in solution.

(5) Separation of Mono- and Dialkyl Sulfates. Mono- and dialkyl sulfates have been separated by solution in butane and partial evaporation of the solvent [98]. This causes separation of the major proportion of the monoalkyl sulfate, which can be removed.

(6) Separation of Sulfates and Sulfonates. For the separation of sulfonates from alkyl sulfates, Swindells [99] and Blank and Kelley [97] made use of the acid stability of the sulfonate. When refluxed in hydrochloric acid solution (25 or 50%), the alkyl sulfates hydrolyze to fatty alcohols. Swindells [99] removed the alcohols by petroleum ether extraction, then reextracted the aqueous acid layer with butanol to isolate the sulfonate. Blank and Kelley [97] added barium chloride solution to the cooled hydrolyzate together with a volume of acetone equal to that of the total solution to aid solubilization of barium alkylarylsulfonate. The desulfated fatty alcohols could be extracted with n-pentane.

(7) Extraction from Nondetergent Matrices. Anionic surfactants may also be found in compositions such as foods, paints, inks, and polishes, often in low concentrations. Extraction from these matrices is often simpler than from detergent products.

If the matrix is insoluble in solvents, then any suitable solvent for the surfactant under examination may be used. Hoffpauir and Kettering [100] determined small amounts of soaps on cotton by Soxhlet extraction with ethanol or isopropanol followed by acidification and extraction of the soaps (as fatty acids) with petroleum ether.

In some cases, especially with proteinaceous materials, the matrix may be made insoluble by suitable treatment, e.g., with acids, alkalis, or alcohol or by heating. Harris and Short [101] used extraction with alcohol or 7 N sodium hydroxide to remove sodium dodecylbenzenesulfonate from canned food, although recoveries were not quantitative.

Gould [102] developed a special extractant to remove anionic surfactant from protein: Magnesium sulfate (2 g) was dissolved in 360 ml of water and 40 ml of 0.1 N sodium hydroxide added. This solution was diluted to 1 liter with acetone. The finely cut sample was treated with 20 ml of the extractant and left to stand for 18 to 24 hr.

Haslam and Willis [103] describe a number of procedures used for the extraction of anionics (alkyl sulfates, alkylsulfonates, alkylarylsulfonates, sulfonated fatty esters) from polymers in paint emulsions (PVC, PTFE, and butadiene/styrene). As a general extracting solvent, methanol was found very useful, but in some cases more extensive treatment was required: (a) If methanol did not break the emulsion, the sample was evaporated to dryness and the resulting gel broken up and then extracted with hot water. (b) Plasticizer present was removed by ether extraction. The extracted sample was dissolved in tetrahydrofuran, the polymer precipitated with alcohol and removed by filtration. The anionic remained in the filtrate and was recovered by evaporation to dryness. (c) In PTFE emulsions containing fluorinated acids and polyethoxylated compounds, these were removed by prior extraction with ether. The insoluble material was treated with water to remove the anionic surfactant.

2. Foaming Extraction

Surface-active substances generally consist of a polar (hydrophilic) and a nonpolar (hydrophobic) hydrocarbon part. At gas-liquid interfaces the molecules orient themselves so that their polar parts are mainly in the liquid phase if it is polar, or their nonpolar parts if the liquid is nonpolar. This adsorption and orientation at the interface occurs because the hydrocarbon part has little affinity for polar solvent molecules and so tends to remove itself from solution by collecting at the interface.

By deliberately creating large gas-liquid interfacial areas in the solution in the form of bubbles, a concentration of surfactant is achieved at the bubble surface. When the bubbles reach the top of the main body of solution, they form a foam which is relatively concentrated in adsorbed material and therefore, if it is removed, a separation of surfactants may be obtained from nonsurfactant material.

a. Factors Affecting Foaming Extraction. Lemlich [104] has presented a comprehensive review of the theory and technique of the various adsorptive bubble separation methods and refers to several applications in the separation of anionic detergents from water [105, 106] and from sewage [107–111].

Foam separation processes are affected by a large number of variables that may be classified as feed (e.g., flow rate, concentration, and nature of surfactant), operating (e.g., air rate and humidity, bubble size, foam reflux) and design (e.g., height and diameter of foam column, height of the foamed liquid above the aerator, height of foam above the solution–foam interface, and the position of feed introduction to the column). The effects of variables on surfactant removal have been discussed by Grieves et al. [112, 113], Kishimoto [114], Pushkarev [115], Rubin and Jorne [116], Rubin and Melech [117], and Brunner and Lemlich [105].

Okabe and Ishii [118] found that the addition of various inorganic salts to waste water greatly improved the removal of sodium dodecylbenzenesulfonate. Presumably the salts acted to lower the solubility of the surfactant and increase surface adsorption.

   b. Apparatus and Procedure for Foaming Extraction. The apparatus for carrying out foam fractionation studies is simple to construct, see e.g., Grieves et al. [113] and Skomoroski [119]. In Skomoroski's design, the fractionation column is made of glass with a diameter of 5 cm and a length of 1 m, although columns of different diameter and length could be used. The column is also designed so that the section containing a fritted glass disperser (nominal pore size 4–5.5 μm), or the column itself, can be replaced. The top portion of the column can also be replaced or extensions added.

The air is drawn from a cylinder, metered with a rotameter and presaturated by passing it through a flask containing water. Before entering the bottom of the column it is passed through a second flask which acts as a liquid trap. A mercury U–tube is inserted between the saturator and trap flasks to act as a safety valve to prevent excessive pressure buildup behind the fritted glass aerator at high flow rates. The column is surrounded by a constant-temperature jacket so that the effect of temperature on foaming can be studied.

The column is filled with a measured volume of liquid to about 15 cm from the top, the solution is aerated, and the foam collected in a receiver via a U–shaped collection tube. Samples of the foamed solution may also be taken from the bottom of the column by means of a tap. In a typical experiment, 3 liters of dodecylbenzenesulfonate (DBS) solution was foamed in a 10-cm diameter column at 24°C. The air flow was 30×10^2 ml/min (STP) for 3 min. A concentration of 8.04 ppm before foaming fell to 2.07 ppm after foaming, a reduction of 74%. Since this was removed in only 30 ml of foam, the actual concentration of DBS in the foam was 600 ppm giving an enrichment factor of 289 times.

   c. Solvent Sublation. In some cases, the separation achieved by bubble fractionation can be greatly increased by placing an immiscible liquid on top of the sample solution. This liquid acts as a trap for the adsorbed material released by the rising bubbles. This method has been called "solvent sublation" [120]. Wickbold [121] made use of this technique for the removal of low concentrations of anionic surfactants from surface waters: One liter of acidified sample solution was placed in a column 50 cm long and 6 cm in diameter. Ethyl acetate (100 ml) was placed on top of the sample. Nitrogen, at 50 to 60 liter/hr, after passing through a presaturator also containing ethyl acetate, was bubbled through the sample for 5 min. The ethyl acetate in the top layer was then drained off and the extraction repeated with a second portion of 100 ml of solvent. The combined extracts were evaporated to dryness and the residue used to determine the amount of anionic present in the sample. Quantitative recovery of added anionic surfactant (0.18 ppm) was obtained.

     d. Ion Flotation. The technique of ion flotation has also been used in the determination of surfactants below their CMC. Tomlinson and Sebba [122] separated potassium oleate (0.1 to 1.5 mg/liter) by adding the cationic dye crystal violet to an alkaline solution of the test sample and aerating for 15 min. The soap/dye complex rises to the surface as an insoluble scum, which can be removed. Lovell and Sebba [123] applied a similar method to separate potassium laurate at concentrations of up to 40 mg/liter.

C. Chromatographic Methods

The extensive growth of chromatographic techniques over the past 25 years has provided the analyst with a range of methods that may be adapted to solve the majority of problems in anionic surfactant analysis. Isolation from commercial preparations, from mixtures of other classes of surfactants or from sewage and waste water is often conveniently carried out by using ion exchange procedures. Separation of individual anionic surfactants from each other and subtle analyses of the components present in a particular manufacturer's product can be obtained by using paper, thin-layer, or adsorption/partition column chromatography.

1. Ion Exchange Chromatography

Separation of detergent mixtures by ion exchange is generally preferable to solvent extraction. The components are separated on the basis of their ionic type rather than their relative solvent solubilities and this results in much cleaner fractionation. In addition, the anionic surfactant may be separated easily from inorganic constituents, a separation that requires great care when solvent extraction is used.

     Much has been published about the use of ion exchange resins for the separation of anionic and other surfactants. Most of the methods make use of resin columns but batch processes (i.e., stirring the resin with the solution under examination) are preferred by some workers. One, two, or three columns of different ion exchangers, or one column of mixed ion exchangers have been used. Neu [124] was one of the first to report the use of a column of ion exchange resin to analyze anionic surfactants. He used the cation exchanger, Wolfatit K, to convert sulfonates to the corresponding sulfonic acids, which were then determined in the effluent by titration with alkali.

Later Takahama and Nishida [125] used Amberlite 120 in a similar process.

a. Separation of Anionic from Nonionic Components. Much of the early work on the use of ion exchange in the analysis of surfactants was devoted to the separation of anionic from nonionic components because this was important for product analysis.

Hempel and Kirschnek [126] were among the first to obtain complete separation of anionics from nonionics by using a column of the polyacrylic acid type cation exchanger Lewatit CN 206 in the acid form. By washing the column with water most of the anionic could be eluted and the rest was recovered by elution with 5% acetic acid solution. After being washed to neutrality with water, the nonionics were recovered by elution with methanol. Normally, it would be expected that nonionic surfactants would not be held on an ion exchanger because they lack ionic groups, nor should anionic surfactants be held on a cation exchanger. Wickbold [127] discusses the abnormal behavior of surfactants in aqueous solution and attributes it to the formation of micelles. However, he could not satisfactorily explain the unexpected adsorption of nonionics observed by Hempel and Kirschnek [126] but considered that micelles played an important part.

The micelles are not broken up by the exchanger, and consequently the surfactant molecules do not penetrate into the resin pores. This results in a very limited exchange capacity since only those sites at the surface of the resin are available for reaction. To avoid this situation, Wickbold [127] suggests firstly replacement of the water by an organic solvent to suppress formation of micelles, and secondly the use of a resin of larger pore size. He found a mixture of methanol and methylene chloride in the ratio 80:20 with the resin Dowex 21 K very suitable.

As a consequence, the use of aqueous solutions for ion exchange of surfactants was gradually phased out and replaced by the use of organic solvents.

At about the same time, Rosen [128] described the use of the strong anion exchange resin Dowex 1 for the separation of anionics and nonionics by a batch technique. Initially the anionics were not recovered from the resin, but subsequently [129] a scheme was devised for the selective elution and separation of different anionics complexed with the resin: 4-5 g of the surfactant mixture, dissolved or dispersed in 100 ml of water, were stirred for 4-5 hr with 20 g of 200-400-mesh Dowex 1-X4 anion exchange resin (chloride form). The resin was filtered and washed several times with 95% ethanol and hexane to remove the nonionics. The anionics were then recovered from the resin by the procedure given in Figure 1.

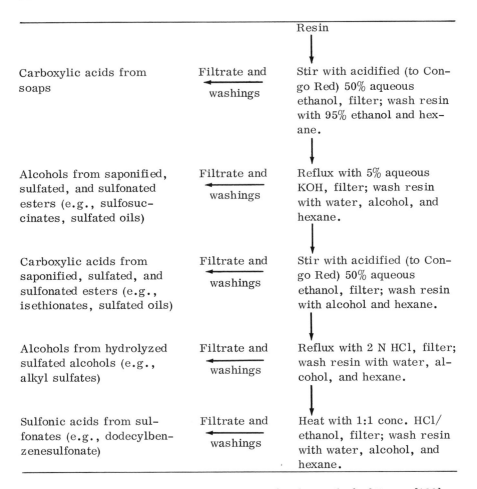

FIG. 1. Separation of anionic components by the method of Rosen [129].

    The recovery of anionics by this procedure was generally only about 80%, probably because of the use of aqueous solvents, but this was quite satisfactory for qualitative identification purposes.

    Ginn and Church [130] obtained almost complete recovery (within 5%) of anionics from nonionics by two-stage column chromatography. The mixture of surfactants in 50% aqueous isopropanol was passed through a strongly acid cation exchanger (Dowex 50-X4) which sorbed all cations and converted the anionic surfactant to its acid form. The effluent from this column (now containing the nonionic and acidified anionic) was passed through an acid-absorbing basic anion exchange resin (Duolite A-7, converted to the hydroxide form). Surfactant anions were sorbed by forming acid salts with the amine functional groups of the resin. Nonionic surfactant was unaffected by the resin and remained in the effluent. The anionic material could then be eluted from the anion exchanger with 2% sodium hydroxide in 50% aqueous isopropanol, and was finally isolated by a salting-out procedure.

Although suitable for the isolation of a wide range of anionic surfactants, the method was found unsuitable for those materials that were unstable to acids and alkalis (e.g., ester sulfonates and sulfates) since these were decomposed either by the resins or the alkaline eluent.

Voogt [131-133] extended this system to three columns in order to remove soaps as fatty acids. An alcohol extract of the sample was passed successively through columns containing (a) Dowex 50-X8(H^+) on which the salts present were converted into the corresponding acids, (b) Dowex 1-X2 (acetate form) on which only the strong (sulfonic or alkyl sulfuric) acids were adsorbed, and (c) Dowex 1-X2(OH^-) on which the fatty acids were adsorbed. The effluent again contained the nonionics. The use of low cross-linked anion exchange resins was found necessary to give quantitative elution of the anionics.

The elution of the anionics from the acetate form anion exchanger was possible either with 0.5 N alcoholic hydrochloric acid or with 0.3 N sodium or ammonium acetate in 75% aqueous alcohol. The acetate eluent was preferred when acid-hydrolyzable components (e.g., fatty alcohol sulfates and ester-type compounds) were present; it also had the advantage that regeneration of the column was not required. However, in order to identify the eluted anionic it was necessary to remove the acetate salt. For materials which are acid stable (e.g., alkylbenzenesulfonates), Voogt suggested the use of cation exchange followed by evaporation until free from acetic acid. With less stable materials (alkyl sulfates, alkoyltaurines) an ion exclusion technique was preferred (see Sec. III.C.5.c), although this gives a fraction containing the anionic in the acid form. The part of his scheme concerning the separation of soap is discussed in Sec. III.C.1.b.

Arpino and de Rosa [134] used only the first two columns of the Voogt scheme for the separation of sodium dodecylbenzenesulfonate from ethoxylated octylphenol, eluting the anionic with 0.5 N methanolic hydrochloric acid.

Wickbold [127] also used a two-column system to separate anionics from nonionics, first a cation exchange resin (any strong acid type was found to be suitable) with the anion exchanger underneath. The anion exchanger, Dowex 21K, was converted to the hydroxide form with dilute sodium hydroxide solution then washed until neutral with methanol:methylene chloride (80:20). The anionics retained on the second column were eluted quantitatively with 15% hydrochloric acid in methanol:methylene chloride, prepared by passing hydrochloric acid gas into the mixture until saturation (about 35%), followed by appropriate dilution. However, since this resulted in partial or complete cleavage of acid-hydrolyzable materials, Wickbold completed the hydrolysis by refluxing the eluent. He then extracted the cleavage products leaving the acid-stable anionics behind, thus separating acid-stable and acid-hydrolyzable anionic surfactants.

To simplify the analysis of anionics and nonionics, Wickbold suggested the use of a bed of mixed resins. The cation and anion exchangers were packed together in a single column before passing the sample of actives through. The effluent from this column contained only the nonionic materials. The mixed bed was then eluted with methanolic hydrochloric acid to remove anionics, as in the two-column system. Using both systems, recoveries of 98-102% of anionics were obtained.

Both Blumer [135] and Czaja and Awerbuch [136] used Wickbold's two-column scheme for the analysis of commercial detergent products. Blumer selected the resins Dowex 50-X1(H^+) and Dowex 1-X1(OH^-), whereas Czaja and Awerbuch used Amberlite IR-120(H^+) and Amberlite IRA-400(OH^- or Cl^-).

Bey [137] recognized the need to extend ion exchange separation methods to newer materials that were coming into common use. His method considered particularly soaps, alkyl sulfates, alkylsulfonates, acylisethionates and taurates, sulfosuccinates and ether sulfates. A two-column arrangement was again used, with a cation exchanger and an anion exchanger in series. The resins used were Dowex 50W-X4(H^+) and Dowex 1-X2(OH^-), respectively. The sample (1 g) was prepared as a 0.5% solution in 96% ethanol. For less-soluble materials Bey suggests the use of 50% ethanol, although urea, if present, was not removed quantitatively by the cation exchanger in this solvent. After the sample had been passed through the columns at 5 ml/min, they were rinsed with a further 100 ml of solvent to wash nonionic material through.

The actual method for elution and identification of anionic material is determined by the nature of the materials in the sample. Bey considers alkyl sulfates and alkylbenzenesulfonates on the one hand and the presence of the remaining anionic types on the other. The types actually present were determined by preliminary thin-layer chromatography (see Sec. III.C.3.a).

If the sulfates or benzenesulfonates were in admixture with toluene- or xylenesulfonates only, the anion exchanger was first eluted with 1 N aqueous hydrochloric acid (250-300 ml) followed by water (100 ml). The eluate contained the short-chain sulfonic acids. The resin was then removed from the column and refluxed with 100 ml of 2 N methanolic hydrochloric acid to hydrolyze alkyl sulfates. After hydrolysis was complete the resin was filtered off and washed with a little petroleum ether followed by methanol. The filtrate was diluted with an equal volume of water and the fatty alcohols were extracted with petroleum ether. The aqueous phase was evaporated to remove methanol, acidified with concentrated hydrochloric acid, and long-chain sulfonates (as sulfonic acids) were extracted with diethyl ether.

When other anionic surfactant materials were also present in the sample, the same procedure was followed as far as the stage involving hydrolysis with methanolic hydrochloric acid. Then the filtrate was neutralized with methanolic sodium hydroxide, and the precipitated sodium chloride removed and washed with a little methanol. The remaining solution was passed through a second pair of ion exchange columns, identical to the first, i.e., Dowex 50W-X4(H^+) in series with Dowex 1-X2(OH^-), and washed through with 100 ml ethanolic solvent.

The eluate now contained nonionic materials resulting from the hydrolysis, including fatty alcohols from alkyl sulfates, short-chain alcohols from sulfosuccinates, alkylphenol and fatty alcohol polyglycol ethers from ether sulfates and, especially, fatty acid methyl esters from acylated materials.

Remaining on the anion exchanger were anionic cleavage products such as long-chain alkyl- or alkylarylsulfonic acids, sulfosuccinic acids, taurines, and isethionic acids. These were eluted with 2 N methanolic hydrochloric acid.

König [138] combined the schemes of Wickbold [127] and Ginn and Church [130] for the separation of anionics, soaps, and nonionics. Three columns were used in series, consisting of (a) the cation exchanger Dowex 50-X4(H^+), (b) the weakly basic anion exchanger Dowex 1-X2(Cl^-), and (c) the strongly basic anion exchanger Dowex 1-X2(OH^-). A solution of the sample in 96% ethanol was passed through and the nonionics were washed out with ethanol. Surfactants of the sulfate and sulfonate type were retained on the second column and could be eluted with 2% sodium hydroxide in 50% aqueous isopropanol. Further identification of the types present was made by the use of hydrolysis reactions with separation of the cleavage products [129, 137]. Carboxylic and aminocarboxylic acids were retained on the third column.

   b. Separation of Soaps from Other Surfactants. The simplest application of ion exchangers to soap analysis was the use of the cation exchanger Dowex 50 in the acid form by Jenkins [139]. Passage of the sample through the resin converted the soap to fatty acids, which were collected and titrated.

Often, however, the soap is present together with synthetic anionic surfactants and some means of separating the two types is required. Although this separation can be performed by merely acidifying a cold aqueous alcoholic solution of the mixture and extracting the soap fatty acids formed with diethyl or petroleum ether, separation can also be achieved by ion exchange.

In general, most of the methods for the separation of soaps rely on the fact that the liberated fatty acids will pass through a weak base (e.g., chloride form) anion exchanger but will be retained by a strongly basic anion exchanger (e.g., hydroxide form).

Thus, in the three-column scheme of Voogt [131-133] a column of Dowex 1-X2(OH^-) was included specifically for the retention of soap fatty acids formed on the cation exchanger and which had passed through the weak base (acetate) anion exchanger. Quantitative elution of the acids was achieved with 250 ml of 0.2 N potassium hydroxide in 70% aqueous ethanol; they were recovered by acidification followed by extraction with light petroleum.

Wickbold [127] varied the order and number of columns according to the nature of the sample mixture. If only a mixture of synthetic anionics and soap was present, a single chloride-form anion exchanger was sufficient to effect separation, for the soap (as the sodium salt) passed through whereas the synthetics were bound. For a soap-nonionic mixture, a two-column scheme was required. The soaps passed first through a cation exchanger in the acid form to free the soap fatty acids which were then passed onto Dowex 21K in the hydroxide form, where they were retained. The nonionics, of course, were found in the effluent from both columns.

Elution of the fatty acids was carried out with 15% hydrochloric acid in methanol:methylene chloride (80:20). After evaporation of most of the acid, aqueous methanol was added and the fatty acids extracted with petroleum ether. However, this procedure converted a significant amount of the acids to their methyl esters, and thus, after evaporation of the petrol, the residue was redissolved in isopropanol and excess sodium hydroxide added to saponify the esters.

When both synthetic anionics and nonionics were present with the soap, Wickbold made use of a three-column scheme, passing the sample successively through a cation exchanger (H^+ form), an anion exchanger (Cl^- form), and an anion exchanger (OH^- form). König [138] used a similar arrangement in his separation scheme but eluted the soap fatty acids from the hydroxide-form exchanger with 2% sodium hydroxide in 50% aqueous isopropanol.

A two-column arrangement, consisting of Dowex 50W-X4(H^+) and Dowex 1-X2(OH^-) in series, was used by Bey [137]. The acids of both anionics and soaps formed on the cation exchanger were retained by the strong base anion exchanger. However, the soap fatty acids could be eluted selectively with 300 ml of 1 N acetic acid in ethanol.

c. Separation of Anionic Types. The main use of ion exchange has been for the separation of anionic surfactants as a group. However, a few instances occur in the literature of the separation of anionics one from another.

Voogt [133] described the separation of sodium isethionate from lauroyl isethionate. In the main scheme, these two materials were bound to the Dowex 1-X2 (acetate form) anion exchanger as the acids, but could be successively eluted with 50 ml aqueous 0.3 N ammonium acetate and with 150 ml 0.3 N ammonium acetate in 75% ethanol, respectively. After cation exchange of the isethionate fraction, isethionic acid and acetic acid were separated by means of ion exclusion.

The separation of components with increasing degrees of sulfonation in alkanesulfonates was performed by Mutter [140]. An alcoholic extract of the sample was passed first through a strongly acid cation exchanger (Biorad AG 50W-X8), then through a weakly basic anion exchanger (DEAE Sephadex A 25, amine form) arranged in series, and then washed with alcohol. The sulfonates were held on the anion exchanger. Alkanedisulfonates and polysulfonates were eluted with 1 N ammonium bicarbonate in water or water-methanol, whereas monosulfonates were eluted with 0.3 N ammonium bicarbonate in water:propanol (40:60 v/v).

d. Separation and Isolation of Anionics from Commercial Preparations. Probably the great majority of analyses with ion exchange have been concerned with commercial detergent preparations; many of these have been discussed in Secs. III.C.1.a and b [124, 125, 127, 130-138]. Several other specific applications have been described.

O'Donnell [141] has given a comprehensive procedure for the analysis of household detergents (light and heavy duty) in which the anionics of interest include alkylarylsulfonates, alkyl sulfates, alkylethoxy sulfates, and soaps. The scheme is based on that of Voogt [133] but only the cation exchange, in this case Dowex AG 50W-X4(H^+), and strong base anion exchange, Dowex AG 1-X2(OH^-), columns were used whereas 50% aqueous isopropanol was substituted as the solvent medium.

The isolation of anionic surfactants from shampoos has been described by Newburger [142, 143] and Gabriel [144]. Newburger used the weakly basic anion exchanger Amberlite CG-45 Type 2 to separate alkyl sulfate from fatty acid alkanolamide and soap. The sample was dissolved in acidified alcohol before passage through the column which was subsequently washed with more acidified alcohol. Alkyl sulfate is bound onto the column, alkanolamide and soap pass through. The anionic was then eluted with an ammoniacal methanol solution of ammonium carbonate and recovered by evaporation (excess carbonate was also removed by this operation). Newburger notes that in a shampoo containing stearyl alcohol, the alcohol also eluted with the sulfate, having apparently dissolved in the aliphatic (hydrophobic) chain of the anionic.

The soap was recovered from the column effluent, after acetone extraction to remove the alkanolamide, by extracting the residue with hot water, acidifying, and finally extracting the soap fatty acids with chloroform.

The method described by Gabriel [144] results from the work of the present authors. The Voogt system was found to be unsatisfactory for shampoos containing amphoteric surfactants which were either destroyed or irreversibly held on the first column (cation exchanger), but reversing the order of the columns provided a more useful system for shampoos. Soaps, if present, were first removed as fatty acids by acidification and solvent extraction of the shampoo nonvolatiles prior to ion exchange. This has a two-fold purpose: first, it avoids possible blockage of columns due to the insolubility of the released fatty acids in the methanol solvent, and second it eliminates the need to separate soaps and amphoterics held on the middle column, Biorad AG 1-X2(OH^-).

A methanolic solution of the extracted nonvolatiles is passed through three columns in series. First through the anion exchanger Zerolit M-IP SRA 151(Cl^-) on which anionic surfactants are held, second through the anion exchanger Biorad AG 1-X2(OH^-) on which the amphoterics are held, and finally through the cation exchanger Biorad AG 50W-X8(H$^+$) which retains cationics and the cation counter-ions of the anionic surfactants. Any nonionics present pass through all three columns and are found in the effluent. The anionic surfactants are eluted from the Zerolit column with 3 N ammonia in ethanol and recovered after evaporation for further analysis and identification.

For the isolation of surfactants from photographic emulsions, Kunimine [145] extracted with ethanol at pH 12-13. The ionic types were further separated by using columns of Sephadex SE(SO$_3$H$^+$) and Sephadex DEAE(OH$^-$), the anionics being retained on the DEAE column. Sephadex apparently showed better sorption and desorption behavior for surfactants than any other ion exchange types (resin or cellulose).

e. Separation from Surface, Underground, and Waste Waters. Several applications of ion exchangers to remove anionic surfactants from waters and sewage have been described. The advantage over other means of isolation (e.g., solvent extraction or adsorption on carbon) is that large volumes of sample may be treated very simply and that the anionic may be recovered, if required, with relatively small amounts of eluting agent. Abrams [146] has used a weak-base anion exchanger (e.g., Duolite A7) for the removal of 1-10 ppm of alkylbenzenesulfonates from laundry wastes and sewage. The surfactant was subsequently recovered quantitatively with 4% aqueous sodium hydroxide solution. In the analysis of sewage, Hughes et al. [147] used Biorad AG 1-X2(Cl$^-$) and eluted the sulfonate with 50 ml of methanolic hydrochloric acid.

Dunning et al. [148] suggested the use of polymeric liquid anion exchangers for the removal of surfactants from sewage. The sample is treated with a kerosine solution of a polyamine with the structure [—N(R)—CH$_2$—X—CH$_2$]$_n$ where R = H, C$_{1-20}$ alkyl, or C$_{6-20}$ aryl; X = a bivalent hydrocarbon radical of dimerized fatty acids; and n = 2-40. At pH 4.78, sodium dodecylbenzenesulfonate at 99 mg/liter was extracted at 99.9% yield.

Riley and Taylor [149] studied the recovery of traces of organics from seawater and found that Teepol (a sec-alkyl sulfate) was fully retained at pH 2 on Amberlite XAD-1 cross-linked polystyrene resin at a concentration of 300 ppb. Full recovery was obtained on elution with 50 ml of 95% ethanol.

f. Mechanism of Anion Exchange Processes. More recently, reports of investigations into the processes involved and the factors affecting the sorption of surfactants on anion exchange resins have appeared in the literature. Wiktorowski and Justat [150, 151] studied the dependence of ion exchange on temperature and the concentration of surfactant for straight and branched-chain dodecylbenzenesulfonates and for dodecyl sulfate with Dowex-1 and -2 resins of various particle sizes. The sorption isotherms were determined at 25, 45, and 58°C, and it was found that the time necessary to reach equilibrium depended not only on the temperature but also on the structures of the ion exchanger and surfactant. The effect of temperature on the concentration of surfactant in the ion exchanger was found to be greater for coarser particles of resin. Adsorption of surfactant on the resin was found to be negligible, except for Dowex 2-X8.

Shamanaev et al. have also made a number of studies [152-154]. Using the porous phosphonic acid cation exchanger KFP-X8(H^+), the absorption of C_{10-16} alkyl sulfates was studied [152] and found not to be dependent on the ion exchange properties of the resin, but due to hydrophobic interaction between the surfactant and the resin matrix. This occurred, however, only in the presence of electrolytes. With univalent electrolytes, the degree of sorption increased with electrolyte concentration and hydrochloric acid was found to exert the strongest influence (compared with lithium and sodium chloride).

Using the anion exchanger AV 17 [153], the dependence of the diffusion coefficient on temperature was studied for C_{10-16} alkyl sulfates and found to exhibit an inflexion point which depends on the chain length. Sample capacities and equilibrium constants for this system were also given [154].

2. Paper Chromatography

Paper chromatography is essentially a simple form of partition chromatography. A sheet of cellular paper (e.g., Whatman No. 1) is spotted with a solution of a small amount of material to be separated; after the spot has dried, the paper is placed in a tank containing a solvent (or mixture of solvents). The solvent travels through the paper by capillarity and as the sample spot is passed the components of the sample partition between the solvent and the water held in the paper fibres. The distribution between the two phases varies for different solutes and hence the sample components travel with the eluting solvent at different rates, giving rise to separation. When adequate separation has been achieved, the chromatogram may be removed from the solvent, dried, and the components visualized with a suitable agent.

In practice, chromatograms are most commonly run with the eluting solvent traveling up (ascending) or down the paper (descending). Occasionally circular chromatography has been used, especially by Franks [178], in which the components are eluted horizontally from the center of the paper.

a. Separation of Anionics from Other Detergent Components. Little has been published on the application of paper chromatography to general detergent analysis. Most workers are concerned with the separation of homologous series and only Drewry [155, 156] has applied the technique to general qualitative analysis.

He separated anionics and soaps from other constituents of commercial detergents including cationic and nonionic surfactants, toluene- and xylene-sulfonates, alkanolamides, and urea. The different materials were identified by a consecutive spraying method. Initially the solvent, tert-butanol: ammonia (sp gr 0.88):water (100:6:6), was used for ascending chromatography on Whatman No. 1 paper, but since development times were long (17-18 hr for 20 cm) the solvent was subsequently changed [156] to ethyl acetate:methanol:ammonia (60:15:5). Development in this solvent required only 2-3 hr. Typical R_f values (i.e., ratio of distance traveled by the substance from the starting line to the total distance traveled by the solvent) obtained in this system were: anionics 0.70, soaps 0.60, quaternary cations 0.60, nonionics 0.95, toluenesulfonates 0.15, xylenesulfonates 0.22, urea 0.24.

The anionics were visualized initially by spraying with pinacryptol yellow (0.05% ethanolic solution), which gave brilliant orange spots under UV light for most substances in this class. Sulfated alkylphenolethoxylates, however, gave violet spots whereas soaps gave pale yellow spots that faded rapidly. Heating the paper at 80°C caused color changes. Alkyl sulfates changed to yellow, ethoxylated alcohol sulfates to blue and taurides, sarcosinates, and sulfosuccinates to yellow or blue. Alkylarylsulfonates and sulfated alkylphenolethoxylates retained their original color.

Subsequent spraying with rhodamine B (0.01% ethanolic solution) and examination by UV light caused all anionics and soaps to appear as brilliant orange spots. Further heating of the paper caused color changes similar to those obtained with pinacryptol yellow. The ethoxylated anionics also gave brown spots (which appeared dark green under UV light) on exposure to iodine vapor and blue-green spots with cobalt thiocyanate reagent.

b. Chromatography of Soaps as Fatty Acids. The analysis of soaps, as such, by paper chromatography has received very little attention, but, on the other hand, the free fatty acids have been widely investigated. This interest has arisen from the presence of fatty acids in many important naturally occurring fats and oils, generally known as "lipids." The separation of the fatty acids from other lipid materials (e.g., glycerol esters, cholesterol esters, phosphatides, etc.) is not strictly relevant here but the separation of the various fatty acid homologues is of more direct interest.

Although the lower fatty acids may be separated by normal paper chromatography, the higher members are hardly resolved by this technique because their physical properties change only slowly with increasing chain length and because they are generally insoluble in water. Hence most workers have used the method of "reversed-phase" chromatography. In this technique the paper is impregnated with a nonpolar stationary phase (usually a hydrocarbon oil) and elution is performed with a polar solvent. The acid components then partition between the stationary and mobile phases according to their relative distribution coefficients, thus giving rise to separation. Generally, the shorter the chain length of an acid, the greater the distance it travels because of its greater affinity for the polar mobile phase.

Reversed-phase paper chromatography has been widely used to separate fatty acids according to chain length (homologous series) and degree of unsaturation. Liquid paraffin has been used by a number of workers to separate the saturated fatty acid homologues. Kopecky [157] used 90%

acetic acid saturated with liquid paraffin as the mobile phase, whereas Chayen and Linday [158] used acetone:water (4:1). Tiwari and Srivastava [159] developed the chromatogram in an atmosphere saturated with acetic acid using acetone:methanol (1:3) as solvent. They also used a mixture of liquid paraffin and cetyl alcohol as the stationary phase, developing the chromatogram again with acetone:methanol (1:3) by a descending technique [160].

Hydrocarbon fractions of high boiling point have been used by Kaufmann et al. [161, 162] and Sliwiok [163]. Kaufmann et al. used a fraction boiling from 190 to 220°C with aqueous acetic acid as the mobile phase (from 7 to 9 parts acetic acid to 3 to 1 parts water). Sliwiok evaluated fractions boiling from 190 to 200°C and 220 to 230°C with mixtures of acetone:acetic acid:water in the ratios 8:4:1 and 8:2:1, respectively, as mobile phases. He found the combination of the 220-230°C fraction with the 8:4:1 solvent ratio to be the most effective. The use of silicone oil was preferred by Mangold, Schlenk, and co-workers [164,165] in conjunction with aqueous acetic acid (85%). Chromatography was performed at 30°C by an ascending technique.

Although reversed-phase chromatography can give excellent separations, the reproducibility can be poor unless precautions are taken. If the stationary and mobile phases show mutual solubility, presaturation is necessary. This is especially important for the mobile phase, otherwise the stationary phase will be stripped from the paper during chromatography. Furthermore, since the separation depends on partition of components between phases, a constant temperature should be maintained. On the other hand, controlled variation of temperature or mobile-phase composition can be utilized to regulate the migration of components to practically useful distances [165].

In order to achieve separation of components over a short migration distance and thus reduce the diffuseness of component spots, Franks [166] applied the technique of gradient elution analysis. He described an apparatus in which the composition of the developing solvent (aqueous acetic acid) could be changed continuously in an exponential manner while chromatography was proceeding. Using liquid paraffin as the stationary phase, separation of saturated or unsaturated acids could be obtained over a distance of 15 to 20 cm, as measured by the solvent front. The method was applied to the separation of long-chain soap fatty acids obtained from acetone extracts of laundered articles. In view of the versatility of this technique for the separation of materials of widely varying polarity, it is surprising that the work by Franks remains the only example of its kind. Table 1 lists the R_f values of saturated fatty acids obtained by the reversed-phase methods described above.

The behavior of unsaturated fatty acids on a reversed-phase system is such that, for the same chain length, the more highly unsaturated members show the greatest migration. In a mixture of naturally occurring fatty acids, differences in both chain length and unsaturation may be encountered. It has been found that, in chromatographic terms, a double bond reduces the

TABLE 1

R_f Values of Fatty Acids by Reversed-Phase Paper Chromatography

Fatty acid and number of carbon atoms	R_f values found by			
	Tiwari [159]	Chayen [158]	Schlenk [165]	Franks [166]
Capric, C_{10}	—	—	—	0.59
Lauric, C_{12}	1.000	—	0.79	0.46
Myristic, C_{14}	0.905	0.55	0.73	0.33
Palmitic, C_{16}	0.750	0.40	0.59	0.21
Stearic, C_{18}	0.656	0.23	0.40	0.13
Arachidic, C_{20}	0.505	—	—	—

effective chain length of an acid by two methylene groups. This gives rise to certain "critical pairs" of acids which migrate similar distances; for example, palmitic acid (C_{16}, saturated) is barely separated from oleic acid (C_{18}, monounsaturated) and myristic acid (C_{14}, saturated) barely separated from linoleic (C_{18}, diunsaturated). If the unsaturated acids are minor components, their detection can be difficult. Such pairs may, however, be separated by such techniques as chromatography before and after hydrogenation [162], by formation of the thiocyanogen derivatives [167, 168], or by the formation of oxygenated derivatives by means of peroxidic solvents [169].

Of the methods developed for the visualization of long-chain fatty acids, perhaps the most widely used have been the procedures involving the formation of heavy metal soaps, especially those of copper [161, 170, 171]. The chromatograms are treated with a solution of cupric acetate, washed to remove excess reagent, then immersed in potassium ferrocyanide solution. Other heavy metals have been investigated by Tiwari [159, 160] including cobalt, nickel, and mercury. Kopecky [157] and Franks [166] have used sulfide reagents to visualize lead and silver soaps, whereas Buchanan [169] used sym-diphenylcarbazide (in 95% ethanol) to identify mercury soaps.

The most common reagent for visualizing unsaturated acids is iodine vapor, which gives brown spots on a white background [157, 164, 165]. This reagent has the advantage of being reversible, the spots fading slowly on standing after removal from the vapor or more quickly on heating.

An interesting method for the visualization of both saturated and unsaturated acids involved the use of α-cyclodextrin (cyclohexaamylose) to form inclusion compounds [164]. Subsequent exposure of the chromatogram to iodine produces white spots for saturated acids and yellow or brown spots for unsaturated acids on a violet background.

Chayen [158] has used the dye Nile Blue under alkaline conditions (triethanolamine) to visualize saturated fatty acids. Churacek et al. [172] suggested chromatography of the acids as their 4-[4-(dimethylamino)phenylazo]-phenacyl esters on paper impregnated with dimethylformamide, using a mixture of light and heavy petroleum ethers as the mobile phase. The advantages are said to be the rapidity, sensitivity, and simplicity of the method, and the easy identification by the use of colored derivatives.

Kaufmann [173] has reviewed the various methods used for visualizing fatty acids after chromatography with regard to quantification. Other useful discussions on the chromatographic analysis of fatty acids, including paper chromatography, may be found in the books by Markley [174], Patai [175], and Gunstone [176].

c. Chromatography of Alkyl Sulfates. The majority of paper chromatographic separations of alkyl sulfates are concerned with their resolution into homologous constituents. Holness and Stone [177] were perhaps the first to effect such an analysis. They separated the n-C_8-C_{18} components by ascending chromatography on Whatman No. 1 paper at 30°C. For the lower members (C_8 to C_{12}), the mobile phase was 15% aqueous ethanol, whereas 40% aqueous ethanol was required for adequate separation of the higher members (C_{14} to C_{18}). Visualization was achieved by spraying with pinacryptol yellow and examining under UV light, when the spots fluoresced bright orange on a pale yellow background.

Since alkyl sulfates are quite soluble in both water and alcohols, normal-phase chromatography would not be expected to give very good separation of homologues. The use of a reversed-phase system should improve the selectivity since a stationary phase can be chosen in which the alkyl sulfates are more or less soluble, according to their chain length. Long-chain fatty alcohols have been found to be suitable materials.

Franks [178] used paper impregnated with cetyl alcohol to investigate the separation of the n-C_{12}-C_{18} alkyl sulfates by both ascending and circular chromatography. Because he used aqueous ethanol as the mobile phase, it was necessary to presaturate with cetyl alcohol. The chromatograms were developed over 10-24 hr, dried, then visualized by immersion in cupric acetate solution followed by spraying with rhodamine 6BG. Components appeared crimson on a pink background in daylight and dark purple on a yellow background under UV light.

The method had the disadvantage, however, like that of Holness and Stone, that no single solvent composition would separate all components. If the ethanol content was too low (e.g., 50%) the higher members hardly migrated; if the ethanol level was increased (e.g., to 62-75%), then, although the higher members began to migrate, the lower homologues traveled with the solvent front (see Table 2).

TABLE 2

R_f Values for the Separation of Homologous Alkyl Sulfates

Alkyl chain	Franks [178]		Borecky [180]		Pueschel [188]
	50% ethanol	75% ethanol	methanol:NH$_3$ (1:1)	methanol:NH$_3$:85% HCOOH (50:50:1)	methanol:NH$_3$ (1:1)
Octyl, C$_8$	—	—	1.00	1.00	0.89
Decyl, C$_{10}$	—	—	0.92	0.78	0.79
Lauryl, C$_{12}$	0.88	1.00	0.82	0.51	0.56
Myristyl, C$_{14}$	0.50	0.90	0.65	0.23	0.30
Cetyl, C$_{16}$	0.11	0.50	0.38	0.07	0.12
Stearyl, C$_{18}$	0.00	0.15	0.16	0.00	0.04

R_f values found by

This problem was overcome by the use of gradient elution [179]. By slowly changing the initial solvent composition (60% aqueous ethanol) by continuous addition of water over about 24 hr good resolution of C_{12}-C_{18} alkyl sulfates was obtained. Care still had to be taken over presaturation of the solvent and to maintain the temperature as constant as possible.

Franks also examined other long-chain alcohols (decyl, lauryl, and tetradecyl) for their suitability as stationary phases, but found that they did not give the same degree of separation as cetyl alcohol.

These methods were also applied to sec-alkyl sulfates (Teepol) and to alkylarylsulfonates. By gradient elution Teepol was fractionated into seven (unidentified) zones. In a mixture of n-alkyl sulfates and alkylarylsulfonates, by using water alone as the mobile phase, the sulfonates migrated whereas the sulfates remained at the origin.

In their work on the identification of organic compounds, Borecky and his co-workers have used paper chromatography quite extensively to separate and identify alkyl sulfates [180-186]. For direct analysis, Whatman No. 3 paper impregnated with 5% lauryl alcohol (in benzene) was used as the stationary phase and methanol:25% aqueous ammonia (1:1) as the mobile phase [180, 181]. The optimum temperature was 35-36°C; at temperatures below 25°C the spots were elongated and the C_{16}-C_{18} sulfates did not migrate. An ethanol:ammonia (4:6) mixture could also be used as mobile phase, but movement through the paper was slower. Incorporation of 1-5% formic or acetic acid into the solvent had the effect of lowering the R_f values obtained (compared with methanol-ammonia alone). Visualization of separated components was by means of 0.05% aqueous pinacryptol yellow and examination under UV light.

In certain cases, alkyl sulfates were separated by chromatography as the alcohols by first hydrolyzing with hydrochloric acid [182, 183]. The alcohols formed were converted to the 3,5-dinitrobenzoates and separated by descending chromatography on paper impregnated with liquid paraffin using formamide:methanol:water (16:1.5:1.5) or dimethylformamide:methanol: water (8:1:1, 4:1:1, or 2:1:1) as the mobile phase. Separated esters could either be detected by direct examination under UV light or by reducing the nitro group to the amino group with stannous chloride, followed by color formation with p-dimethylaminobenzaldehyde (Ehrlich's reagent). Borecky has reviewed the application of paper chromatography to the analysis of anionic surfactants [184-186].

Sewell [187] has used a similar method for the separation of Teepol and sodium lauryl sulfate. The 3,5-dinitrobenzoate esters were separated on plain paper using dry benzene as the mobile phase. After chromatography the paper was sprayed with hydroxylamine hydrochloride, dried, then sprayed with cold alcoholic sodium hydroxide, and dried again. Yellow spots at the origin and the solvent front indicate Teepol and lauryl sulfate, respectively.

Pueschel and Prescher [188] investigated the composition of straight-chain olefin sulfation mixtures using the reversed-phase system of Borecky

[180, 181]. n-Alkyl sulfate reference chromatograms were prepared and the sulfation mixtures were separated using the system 5% lauryl alcohol/ methanol:ammonia (1:1), but by ascending chromatography at room temperature over about 16 hr. For the separation of positional isomers (e.g., the 1,2,3,5, and 8 isomers of sodium hexadecyl sulfate), differentiation was improved by using the system 2.5% lauryl alcohol/methanol:ammonia (3:2). When pinacryptol yellow was used to detect the spots, it was also possible to distinguish between the 1,2 and the 3,5 and 8 isomers by the different colors produced. The lower isomers gave yellow spots, the higher members orange; the difference was thought to be due to the shortening of the main carbon chain.

    d. <u>Chromatography of Sulfonates</u>. Generally, the chromatographic systems used to characterize alkyl sulfates have also been used for sulfonate-type materials. Borecky has investigated the behavior of arylsulfonates as the sulfonic acids using both normal [189, 190] and reversed-phase [190] methods. The "short-chain" sulfonates, such as those of benzene, toluene, xylene, and naphthalene, were separated on Whatman No. 3 paper using lower aliphatic alcohols and ammonia as the developing solvents, especially n-propanol:ammonia [2:1]. After drying at 100°C, the spots were visualized by spraying with 0.05% aqueous pinacryptol yellow and examined under UV light. The characteristic fluorescent colors of compounds with similar R_f values were a further useful guide to identification, e.g., p-toluenesulfonic acid, R_f 0.80, orange-brown fluorescence; xylenesulfonic acid, R_f 0.81, light yellow fluorescence; 1-naphthalenesulfonic acid, R_f 0.80, brownish-yellow fluorescence.

    The true anionic sulfonates were preferably separated by using the reversed-phase system 5% lauryl alcohol/methanol:ammonia:formic acid (50:50:1). Good separations were obtained at 35°C using a descending technique.

    The alkylarylsulfonates could also be separated and identified by conversion to the corresponding phenols or naphthols [191, 192]. A small amount of sample (50 mg) was heated with potassium hydroxide at 360–370°C for 5 min, cooled, and dissolved in water, or alternatively heated with colorless hydriodic acid in a sealed tube at 145°C for 30 min. The products were then treated with excess sodium thiosulfate and the liberated phenols extracted with benzene. For chromatography, the phenols were separated on paper impregnated with formamide or dimethylformamide using hexane or benzene as the mobile phase. Visualization was effected with freshly prepared ferric chloride and potassium ferrocyanide solutions.

    Analysis of lignin sulfonates was carried out by means of chromatographic identification of vanillin, which was formed by thermal decomposition of the lignin sulfonates with sodium hydroxide and nitrobenzene [193].

    Pueschel and Prescher [188] investigated the composition of sulfonation mixtures using Borecky's reversed-phase system. They examined the chromatographic behavior of the sodium salts of C_{10}-C_{18} alkane-, alkene-,

2- and 3- hydroxyalkane-, and 2- and 3-oxoalkanesulfonates but at room temperature and by an ascending method. Generally, the system 5% lauryl alcohol/methanol:ammonia (1:1) could be used quite successfully for separation of homologous series or for separation of mixtures, but in some cases it was found better to make modifications, for example, (a) improved R_f values were obtained for C_{18} alkane- and alkenesulfonates by using methanol:ammonia (3:2) as mobile phase and also by using paper impregnated with 1% lauryl alcohol (in benzene), and (b) greater impregnation was better for C_8-C_{10} sulfonates, and a level of 20% lauryl alcohol was used (Table 3). The various colors developed under a UV lamp after spraying with 0.05% aqueous pinacryptol yellow were a further aid to the identification of components in mixtures.

Later, Pueschel and Todorov examined the behavior of sodium alkylarylsulfonates using 5% lauryl alcohol/methanol:ammonia (3:2) by descending chromatography [194]. Alkyl groups containing heteroatoms were also investigated. They concluded that for one heteroatom present, the R_f values increased along the series S, O, NH (see Table 3).

Aliphatic sulfonates, including taurine and isethionate, were investigated as their ammonium salts by Coyne and Maw in seven solvent systems [195]. Since most of the compounds examined were short-chain sulfonates, normal-phase ascending chromatography at 20°C on Whatman No. 1 paper was used. Washing the paper beforehand in acetic acid greatly improved the definition of the spots subsequently obtained and also eliminated background coloration.

A study was also made of various reagents for visualizing the spots including acid-base indicators, barium chloride-sodium rhodizonate, pinacryptol yellow, and silver fluoresceinate and heating at 80°C (for solvents containing phenol or mesityl oxide) or 120°C (all other solvents). Coyne and Maw found that the most suitable visualizing agents for use with the aliphatic ammonium sulfonates were bromocresol green (0.5% in ethanol) and silver fluoresceinate (one part of 10% aqueous silver nitrate and five parts of 0.2% sodium fluoresceinate in ethanol, prepared immediately before use). With the latter reagent spots appeared pale yellow on a salmon-pink background but darkened on exposure to light. Although pinacryptol yellow is perhaps the most widely used reagent for sulfonates, Coyne and Maw found that it was not sufficiently sensitive for routine detection.

Franks [178, 179] applied his reversed-phase systems (i.e., with and without gradient elution) to the analysis of alkylarylsulfonates claiming good results, although no data are given.

e. Chromatography of Other Anionics. Although the majority of paper chromatographic analyses are concerned with soap fatty acids, alkyl sulfates, and aliphatic and aromatic sulfonates, separations applied to other anionic surfactant types have appeared in the literature.

Drewry [155, 156] separated various amino-group-containing anionics (taurides, N-methyltaurides, and sarcosinates), after acid hydrolysis to

TABLE 3

R_f Values for the Separation of Various Sulfonates

Borecky [189] n-Propanol:ammonia (1:1) Sulfonic acid	R_f	Borecky [190] Butanol saturated with ammonia, R_f	Pueschel [188] Sodium n-alkane-sulfonate	R_f	Pueschel [188] 5% Lauryl alcohol/methanol:ammonia (1:1) Sodium n-alkene-2-sulfonate	R_f	Pueschel [194] Sodium alkylbenzenesulfonate	R_f
Benzene—	0.74	0.36	Octan—, C_8	0.85			n-$C_{10}H_{21}$—	0.46
p-Toluene—	0.80	0.42	Decan—, C_{10}	0.75			n-$C_{12}H_{25}$—	0.18
Xylene—	0.81	0.52	Dodecan—, C_{12}	0.53	Dodecen—	0.74	n-$C_{14}H_{25}$—	0.07
1-Naphthalene—	0.80	0.49	Tetradecan—, C_{14}	0.25	Tetradecen—	0.46	n-$C_{11}H_{23}$—S—	0.28
2-Naphthalene—	0.82	0.48	Hexadecan—, C_{16}	0.08	Hexadecen—	0.20	n-$C_{11}H_{23}$—O—	0.38
			Octadecan—, C_{18}	0.03	Octadecen—	0.05	n-$C_{11}H_{23}$—NH—	0.51

free the amines, using ethyl acetate:methanol:concentrated ammonia solution (45:45:10) on Whatman No. 1 paper. The components were visualized by spraying with ninhydrin solution (0.2 g in 95 ml of methanol and 5 ml of 2 N acetic acid) and heating at 80°C for 5 min. Taurine (R_f 0.29) gave a purple spot, sarcosine (R_f 0.29) a light purple spot. Further heating at 120°C for 10 min deepened the color of the spots already formed and gave a purple spot for N-methyltaurine (R_f 0.44).

Sulfosuccinates have been characterized by Borecky [182, 183, 185, 186] by hydrolysis to the corresponding alcohols and chromatography of the 3,5-dinitrobenzoate esters. A reversed-phase system was used with paper impregnated with paraffin oil as the stationary phase and methanol:dimethylformamide:water (8:1:1) as the mobile phase. Spots were visualized using stannous chloride followed by Ehrlich's reagent (p-dimethylaminobenzaldehyde).

Various phosphate species, including mono(2-ethylhexyl)phosphate, bis(2-ethylhexyl)phosphate, and orthophosphoric acid, have been separated by Kuzin et al. [196]. They used butanol:methanol:1.5 N ammonia (4:1:1) and butyl acetate:acetic acid:water (13:13:4) as solvents. Visualization was carried out with a solution containing 7 ml of 42% perchloric acid, 10 ml of 0.1 M hydrochloric acid, and 25 ml of 4% ammonium molybdate in 100 ml of water.

3. Thin-Layer Chromatography

Thin-layer chromatography (TLC) may be considered as an extension of paper chromatography in which the separation occurs on a support other than paper. Essentially, a thin layer of powdered support is deposited onto a smooth inert surface, glass plates being most widely used although aluminum and plastic backings are available and are becoming popular.

Generally the support is a strong adsorbent (e.g., silica or alumina) and the mechanism of separation depends on the reversible adsorption of the solute molecules at the adsorbent surface. However, by impregnation of the adsorbent with a suitable stationary phase, it is also possible to perform normal and reversed-phase partition chromatography. This ability to use both adsorption and partition modes gives TLC much greater selectivity than paper chromatography and it is possible to separate components differing only slightly in structure or configuration.

The speed of analysis, too, is generally greater, probably because of the faster rate of travel of solvents through a fine powder layer as compared to a fibrous paper sheet. This has the further advantage that the spots in TLC are smaller and less diffuse than in paper chromatography and, since they therefore contain components at higher concentrations, they are more easily detected. Nonvolatile organic compounds can also be visualized by the use of a general charring reagent such as chromic/sulfuric acid without physically destroying the chromatogram which would occur with a paper chromatogram. Additional information on the nature of various compounds may be obtained, depending on whether spots appear before or after heating and on the temperature reached during heating.

For a fuller discussion on the technique, equipment, adsorbents, and applications of TLC, the comprehehsive treatise by Stahl [197] is recommended.

a. Separation of Anionics from Other Detergent Components. The earliest use of TLC for the separation of anionic surfactants was reported by Mangold and Kammereck [198] in their work on the analysis of industrial aliphatic lipids containing nitrogen, phosphorus, and sulfur. A series of solvents, increasing in polarity, was used to separate classes of lipids on Silica Gel G (silicic acid) plates. Strongly acid sulfates, sulfonates, phosphates, and phosphonates, however, had to be chromatographed on plates modified to contain 10% by weight of ammonium sulfate using the polar solvents chloroform:methanol containing 5% 0.1 N aqueous sulfuric acid (97:3 or 80:20). The analysis time was about 40 min for the front to travel 12 cm.

Complete fractionations were obtained of materials such as primary, secondary, and tertiary amines, amides (including alkanolamides), and quaternary ammonium salts. With the anionics, resolution was obtained for mixtures of N-acylated sarcosine, oleic acid ester of hydroxysulfonic (isethionic) acid, and N-acylated short-chain amino acid (taurine) using the first solvent combination (97:3) and for alkyl sulfates, sulfonates, phosphates, and phosphonates using the second solvent (80:20). Turkey red and monopol oils (sulfated and sulfonated castor oils) were also analyzed for their degree of sulfation or sulfonation using the second solvent. The spots were visualized on the plates by the use of chromic/sulfuric acid.

After class separation, Mangold removed the spots from the plate and eluted them from the adsorbent using a suitable solvent. They were then used for further fractionation into homologues by paper chromatography (see Sec. III.C.2).

The speed of analysis given by TLC encouraged Desmond and Borden [199] to investigate detergent formulations by this technique. They used Aluminum Oxide G as adsorbent with isopropanol as solvent for analysis of the alcohol-soluble fraction of the sample. After being developed for a distance of 10 cm, the plate was dried and spots visualized by a sequence of reagents based on those of Drewry [155, 156]. First, pinacryptol yellow was used and the plate examined by UV light. This was followed by exposure to iodine confirming the position of nonfluorescent spots, especially alkanolamides, and aiding their identification, since the spot for dialkanolamide faded on standing, whereas that for monoalkanolamide faded only slowly. Finally the plate was sprayed with cobalt thiocyanate reagent to characterize ethoxylates and amine oxides. Since anionics tended to have low R_f values with this system (0-0.2), identification was made mainly by the colors given with pinacryptol yellow, which were:

alkylarylsulfonates	yellow
soaps	blue
xylenesulfonates	orange-yellow
toluenesulfonates	orange-red
sulfated alcohols	pale blue
alkylphenolethoxylates	pale blue

Commercial products containing surfactants are often complex mixtures and Arpino and de Rosa [200] concluded, after study of many individual common types, that their separation on a single chromatoplate would be difficult. Therefore they favored a preliminary separation of the ionic types by the use of ion exchange resins [134] (see Sec. III.C.1).

Anionics, as the acids, together with nonionics, could be obtained after passage of an ethanolic solution through a column of Dowex 50-X8(H^+). A sample of the effluent from this column was then subjected to TLC on Silica Gel G plates using petroleum ether (bp 60-70°C):diethyl ether:glacial acetic acid (70:30:2). After development, the spots due to anionics were visualized using 2',7'-dichlorofluorescein and examined under UV light.

Soaps (as fatty acids) gave a principal spot with an R_f of 0.6; sometimes with two other small spots of higher R_f values. Alkyl and ethoxylated alkylsulfates remained at the origin but sometimes gave a spot due to nonsulfated or nonethoxylated alcohols ($R_f = 0.45$). Sulfonates of all types also remained at the origin but gave a weak spot on the solvent front due to nonsulfated material.

Anionics were also investigated after being passed through a cation and anion exchanger in series. After elution from the anion exchanger the anionics were hydrolyzed with hydrochloric acid solution.

The hydrolysis products were chromatographed on Silica Gel G using, respectively, the previously mentioned petrol:ether:methanol (70:30:2) mixture and chloroform:methanol (80:20) saturated with water. On visualization with iodine, the first solvent revealed the presence of soap fatty acids ($R_f = 0.3$). The second solvent was more useful for sulfonated and ethoxylated compounds. Sulfonates gave an elongated spot at $R_f = 0.3$ together with a small spot on the solvent front due to nonsulfonated material. Ethoxylated alcohols from ethoxylated sulfates also gave an elongated spot, but at $R_f = 0.7$, together with two smaller spots underneath.

Conversion of anionics to the acid form before TLC was carried out by Löser [201] because the wide variety of cations used to neutralize the anionic acids commercially caused changes in the R_f values even for the same anions and made identification difficult. Passage of an alcoholic extract of the sample through the cation exchanger Dowex 50W (H^+) ensured that all anionics contained the same cation, hydrogen.

The chromatographic analysis was then carried out in two parts. In the first part, anionics and nonionics were separated on Silica Gel G using chloroform:methanol + 5% of 0.1 N sulfuric acid (97:20); in this solvent the nonionics migrate while the anionics remain near the start. The plate could, at this point, be sprayed with Dragendorff's reagent followed by rhodamine B to identify the types of materials present. But since this solvent did not separate alkyl sulfates, alkylarylsulfonates, and ethoxylated alkyl sulfates from one another, in the second part of the analysis a more concentrated sample was run again on a second silica plate in the first solvent. The lower part of the plate was then sprayed with rhodamine B to determine the position of the anionic components which were scraped off into a small flask and refluxed with 6 N hydrochloric acid for 1 hr. After ether extraction of the hydrolysis products, the extract was evaporated to dryness and redissolved in 0.1 ml of 95% ethanol. Of this solution, 10 μl were then chromatographed on another silica plate for a distance of 4 cm using the original solvent. The plate was removed from the solvent, dried, and placed in a tank containing isopropyl ether and developed for 10 cm. Separated spots due to fatty alcohols, ethoxylated fatty alcohols, and alkylarylsulfonic acids were visualized by spraying with 50% aqueous sulfuric acid and charring at 180°C. The method was applied successfully to a number of commercial washing and cleansing products.

The work of linking ion exchange with TLC was continued by Bey [202, 203] and König [138]. Bey, however, set out with the intention of using TLC before ion exchange so that with a knowledge of the anionics present, he could select suitable conditions for their elution from the resin [137].

Investigations were made of the behavior of various anionics (and nonionics) using alumina and silica for adsorption TLC and 2% dodecanol on kieselguhr for reversed-phase partition TLC. Although some useful separations were obtained, Bey felt that no system was completely satisfactory for the examination of products. However, the use of a two-dimensional method on Silica Gel G gave the desired results. By running in one direction with propanol:chloroform:methanol:10 N ammonia (10:10:5:2) for 70 min, a separation of anionics from nonionics could be obtained (R_f for anionics was 0.61 maximum, for nonionics 0.84 minimum). In the second direction, at right angles to the first run, ethyl acetate:methanol:ammonia (45:2.5:5) was used to separate and identify individual components. In this solvent the anionics migrate very little, whereas nonionics move reasonable distances. Ethoxylated compounds, especially, are further resolved into a series of spots due to the homologues present (although highly ethoxylated materials may remain at the origin). Hence, to improve the distinction between dodecylbenzenesulfonate, alkyl sulfate, and ethoxylated alkyl sulfate a hydrolysis procedure was inserted and carried out directly on the plate; after application of the sample, the bottom of the plate was carefully dipped in concentrated hydrochloric acid, which was allowed to rise 3 cm, and then heated for 1 hr at 90°C. Chromatography in the second direction took only 25 min. Identification of anionics was further aided by noting the colors produced under UV light after spraying with pinacryptol yellow. The nonionics were characterized by using modified Dragendorff's reagent which generally gave white spots, although ethoxylated compounds produced red-brown colors.

Thus an acceptable qualitative analysis of a product for anionic and nonionic constituents could be performed in about three hours.

Two-dimensional TLC was also used by Daradics [204] to separate anionics (dodecylbenzenesulfonates, fatty alcohol sulfates) from nonionics (ethoxylated alkylphenols, ethoxylated fatty alcohols). Silica Gel G was employed as adsorbent with chloroform:methanol:water (80:19:1) as the first solvent and water-saturated methyl ethyl ketone as the second.

In König's ion-exchange/TLC scheme, TLC was considered to be a suitable means of further separating and identifying anionics after their elution from the anion-exchange column. He investigated the suitability of the methods developed by Mangold and Kammereck [198] and Bey [202] for the separation of some 27 commercial anionic surfactants and concluded (as Bey had previously) that separation of all anionic types from one another would not be possible using one-dimensional analysis. However, the methods were suitable for certain types; thus anionics based on fatty acids could be reasonably separated by the Mangold and Kammereck system, whereas a partial separation of sulfur-containing anionics could be achieved using Bey's method [i.e., with the solvent propanol:chloroform:methanol:10 N ammonia (10:10:5:2)]. Bey's method was again best to use when phosphated anionics were also present since these migrated ahead of the other anionic types.

Among other reported applications, Yoshikawa et al. [205] used silica plates and the solvent system methyl ethyl ketone:benzene:ethanol:water to separate anionics from nonionics, urea, alkanolamides, and amines in commercial liquid shampoos and dishwashing formulations. Egginger and Weidauer [206] separated textile auxiliaries (mainly alkyl sulfates and nonionics) using Silica Gel G and ethyl acetate:pyridine (40:60) and found that the R_f values obtained together with characteristic colors in UV light after spraying with pinacryptol yellow gave a reliable identification of components.

Takagi and Fukuzumi [207] investigated 17 surfactants on silica using chloroform:methanol:water and found that only the free (acid) components, not the salts, gave sharp spots. Anionics were also stated to give a peculiar fluorescence when examined under UV light after spraying with sulfuric acid and heating.

   b. Chromatography of Soaps as Fatty Acids. As with paper chromatography, the analysis of fatty acids by TLC has produced a wealth of methods, mostly biased toward the characterization of lipids. Again, we are giving here only a brief outline of methods which could be useful in the analysis of soaps. There are also many useful review articles, notably by

Markley [174], Mangold [208, 209], Williams [210], James [211], and Nichols [212].

As was found with paper chromatography, the similarity in physical properties of the saturated fatty acids results in poor resolution of homologues by direct adsorption TLC [213]. Singh and Gershbein [214], however, published a direct method involving separation on Silica Gel G, using n-butanol saturated with water. The acids were well-resolved and a linear relationship was established between the R_f value and the number of carbon atoms in the acid (see Table 4). The quality of this analysis, together with the fact that a water-saturated solvent was used, suggests perhaps that either the silica is partially deactivated by the solvent, weakening the adsorption of the acids, or that in fact the silica is completely deactivated and the water acts as the stationary phase in a partition system.

Superior separations of acids according to chain length or degree of unsaturation may be achieved with reversed-phase partition TLC using a

TABLE 4

R_f Values for the Separation of Fatty Acids

Fatty acid	Singh [214]	Malins [215][a]	Kaufmann [219]	Hammonds [222][a]	Heusser [216]	Paulose [227][a]
Saturated						
Capric	0.72		0.83		0.45	
Lauric	0.59	0.75	0.77	0.62	0.34	0.53
Myristic	0.45	0.56	0.64	0.49	0.25	0.47
Palmitic	0.33	0.39	0.49	0.35	0.18	0.41
Stearic	0.19	0.23	0.36	0.24	0.11	0.34
Arachidic			0.25			0.28
Behenic						0.21
Unsaturated						
Oleic		0.37	0.51	0.39		0.53
Linoleic		0.57	0.64	0.56		0.61
Linolenic		0.75	0.75	0.69		0.70
Erucic			0.24			0.41

[a] As methyl esters.

nonpolar stationary phase and developing with a polar mobile phase. As with reversed-phase paper chromatography, under these conditions the saturated fatty acids separate according to polarity, i.e., molecular weight, those of highest polarity (lowest molecular weight) migrating fastest. With unsaturated fatty acids of the same chain length, separation is by degree of unsaturation, the more highly unsaturated acids migrating fastest.

Suitable hydrophobic plates for reversed-phase TLC can be obtained by coating silica gel or kieselguhr with silicone or hydrocarbon oils. Malins and Mangold [215] coated thin layers of Silica Gel G by slowly immersing the plates at room temperature into a 5% solution of silicone in diethyl ether. Roomi et al. [213] preferred the technique of allowing the ethereal silicone solution to ascend the plate in a developing tank. This procedure had the advantages of requiring smaller volumes of solution and avoiding the loosening of silica gel from the plate when the temperature of plate and solution were not the same or if the immersion was performed too quickly [208].

To overcome the difficulties of reproducibly coating silica plates (which leads to nonreproduciliby of chromatograms), Heusser [216] made use of the free hydroxyl groups of the silica by reacting them with dichlorodimethylsilane to give silanized silica gel.

Kaufmann and co-workers [217-221] and Hammonds and Shone [222] preferred to use a hydrocarbon oil such as undecane or liquid paraffin as stationary phases. Plates were again prepared by a dipping technique, using a solution of the hydrocarbon (5 or 10%) in petroleum ether. A disadvantage of undecane as stationary phase is that it evaporates slowly and may therefore cause changes in reproducibility of chromatograms.

The use of polyethylene powder as a successful reversed-phase partition medium for fatty acids was noted by Mangold [208], although no results were given.

Suitable solvent systems that have been used for reversed-phase TLC of fatty acids are given in Table 5. Generally, it is necessary to equilibrate the solvent with the stationary phase used to prevent stripping during the chromatographic run.

When a mixture of saturated and unsaturated fatty acids is separated by reversed-phase TLC, the problem of the running together of "critical pairs" occurs again (see Sec. III.C.2.b). Such pairs may be separated by methods similar to those described previously, namely low-temperature TLC [215], chromatography in peroxidic solvents [215], or by bromination or hydrogenation before chromatography, reactions conveniently performed on the plate [220, 221], or by one of the techniques described below.

Hammonds and Shone [222] effected separation of "critical pairs" as their methyl esters, by a combination of a suitable choice of solvent and differential visualization. Using the system nitromethane:acetonitrile:acetic acid (75:10:10)/undecane, the chromatogram was developed over a distance of 10 cm. The developed chromatogram was dried and the resolved compounds detected by spraying with a saturated aqueous solution of ferric chloride

TABLE 5

Solvent Systems for Reversed-Phase TLC of Fatty Acids

Solvents	Composition (v/v)	Reference
Acetic acid:water	3:1 and 17:3	Mangold [208]
		Malins [215]
Acetonitrile:acetic acid:water	70:10:25	Malins [215]
	70:10:20	Roomi [213]
Dioxan:water:formic acid	60:35:5	Heusser [216]
Acetic acid:acetonitrile	1:1	Kaufmann [217]
		Kaufmann [219]
		Kaufmann [221]
	6:4	Kaufmann [220]
Nitromethane:acetonitrile:acetic acid	75:10:10	Hammonds [222]
Acetone:ethanol:water	6:1:3	Mangold [208]

followed immediately by a 0.1 M aqueous solution of sodium molybdate, and then heating at 140°C for 3-5 min. The saturated methyl esters gave orange spots and the unsaturated esters gave blue-purple spots on a brown background. By this means methyl linolenate was completely separated from laurate, and linoleate from myristate. Complete separation of oleate from palmitate was not quite achieved.

Saturated and unsaturated fatty acids, as their methyl esters, can also be separated by adsorption TLC on Silica Gel G after reaction with mercuric acetate to form addition compounds across the unsaturated double bonds [208, 223]. To obtain optimum resolution two solvent systems were applied consecutively. First the saturated methyl esters were separated from the acetoxymercurimethoxy derivatives of all unsaturated esters with a mixture of petroleum ether (bp 60-70°C):diethyl ether (4:1). The derivatives of the unsaturated esters were then separated, in the same direction, according to their degree of unsaturation and regardless of chain length, with n-propanol:glacial acetic acid (100:1). Compounds were visualized with 0.1% diphenylcarbazone in 95% ethanol to give purple spots on a light rose background.

After isolation by TLC, the original unsaturated esters (mono, di, tri, and higher) could be recovered by scraping off the spots and treating these with methanolic hydrochloric acid. The saturated esters could be located by exposure to iodine vapor and removed from scraped spots by extraction with diethyl ether for further analysis.

Chromatography on silica gel impregnated with silver nitrate (5% by wt) provides another alternative for the separation of fatty acids according to their degree of unsaturation, and depends upon the weak complex formed between silver ions and the π electrons of double bonds [224, 225]. The separations achieved, using hexane:diethyl ether (95:5), are similar to those obtained with the mercuric acetate adducts, but the system has the advantage that the adduct-forming reaction step is eliminated, as is the ester regeneration step.

Since both the mercuric acetate and silver nitrate methods separate fatty acids only according to their degree of unsaturation regardless of chain length, they must generally be used in conjunction with other techniques, e.g., gas chromatography, to obtain complete structural information. To simplify and speed this total analysis, Bergelson et al. [226] and Paulose [227] have described methods using only thin-layer chromatography.

Both methods are based on the combination of reversed-phase TLC with silver nitrate impregnation. In Bergelson's method two-dimensional TLC was used. The sample was first subjected to chromatography on silica gel impregnated with dodecane (10% v/v in hexane) using acetonitrile:acetone (1:1) as developing solvent. This led to separation into groups according to the number of carbon atoms. The plates were then impregnated with a solution of silver nitrate (20% w/v) and developed in the second direction with dipropyl ether:hexane (2:3) to achieve separation of the groups into individual components according to their degree of unsaturation. Visualization was by means of alkaline or ammoniacal bromothymol blue.

In the method of Paulose, the silica gel plate was impregnated with silicone oil (5% v/v in petroleum ether), then sprayed with a 10% solution of silver nitrate in 50% aqueous ethanol and dried. The fatty acid methyl esters were developed on this plate using 95% v/v aqueous methanol saturated with both silicone oil and silver nitrate and detected by heating after spraying with ethanolic phosphomolybdic acid.

A separation was achieved of the various unsaturated esters of the same chain length from saturated esters differing in chain length by two carbon atoms, i.e., the common "critical pairs." However, overlap now occurred between other pairs of esters, namely linolenate and ricinoleate, oleate and laurate, elaidate and myristate, and erucate and palmitate although these pairs may be separated by ordinary reversed-phase TLC on a silicone stationary phase [215].

c. Chromatography of Sulfates. Methods for the separation of sulfates from other detergent components by TLC have been discussed above in subsection a, but surprisingly very few workers have attempted to use this technique to separate the homologous series which often occur with these surfactants, especially considering the separations that have been performed using paper chromatography (see Sec. III.C.2.c).

Breyer et al. [228] have made the only noteworthy systematic study of the parameters affecting the reversed-phase TLC of ethoxylated alkyl sulfates of the general formula:

$$R_1-[OCH_2C(R_2)H]_x-OSO_3Na$$

where

$$R_1 = C_{12}-C_{18} \text{ alkyl}$$
$$R_2 = H, CH_3, C_2H_5$$
$$x = 1 \text{ or } 2$$

Separations were designed on the basis of variation in the R_f values, spot shapes, and areas of the surfactants with the variables: (a) composition of the solvent system methanol:28% ammonia; (b) carbon chain length and concentration of the alkanol reversed-phase impregnating agent; (c) type of adsorbent; (d) temperature; and (e) humidity.

From these studies, the best separations were obtained on commercial $250\mu m$ layer glass plates of Silica Gel G, Alumina H, or Alumina G impregnated with a 3-5% (v/v) solution of n-dodecanol in ethanol at 15-30°C and in tanks preequilibrated with the solvent methanol:ammonia (3:2). The use of aqueous pinacryptol yellow (0.05% w/v) together with UV viewing was found to be the most satisfactory spot-detection method. Reference was also made to the application of this system to two other series of anionic surfactants, although no details were given.

d. Chromatography of Sulfonates. Several TLC methods have been described for the analysis of technical sulfonate products. Both Mutter [140] and Küpfer et al. [94] have used TLC in their alkanesulfonate analysis schemes to monitor the species present (alkane mono-, di-, and polysulfonates and unsulfonated alkanes) and to check the selectivity of their various extraction methods.

Mutter used a system similar to that of Mangold and Kammereck [198]; samples were developed on a layer of Silica Gel G deactivated with ammonium sulfate using a mixture of chloroform:methanol:0.1 N sulfuric acid (70:30:6). The components were visualized by exposure to iodine vapor followed by heating to 350°C and appeared as black spots on a white background; R_f values increased in the order polysulfonate, disulfonate, monosulfonate, and unsulfonated matter.

Küpfer effected separation of these components on unmodified Silica Gel G by using n-propanol:concentrated ammonia (7:3) as solvent. Components appeared as yellow spots on a brown background (green spots under UV light) after spraying with 2.5% ethanolic 4,5-dichlorofluorescein with the following R_f values: monosulfonate 0.55, disulfonate 0.36, polysulfonate 0.32. The separation was independent of the carbon number of the alkyl chain and of the number of sulfonate groups.

Mutter and Han [229] have also described the separation and quantification of different types of sulfonates including sodium alkylbenzenesulfonates, sodium alkenesulfonates, sodium hydroxyalkanesulfonates, and sodium alkanesulfonates. Again silica gel modified with ammonium sulfate was used with the solvent system chloroform:methanol:0.1 N sulfuric acid (70:30:6). After the developed plates had been dried, they were placed in a chamber containing sulfuryl chloride and water for 20 min and then heated to 300°C. This process caused charring of the component spots. Quantification was carried out by scraping off the charred spots, mixing with more silica gel and reading the whiteness level with a leukometer. Galvanometer deflection was found to be linearly related to the spot load for known amounts. The calibration curves so derived were then used to analyze unknown samples with an accuracy of 5%.

Identification of short-chain alkylbenzene- or naphthalenesulfonates, commonly used as hydrotropes in detergent formulations, has been described by Dunn and Robson [230] using reversed-phase TLC. Plates of Silica Gel G were impregnated with a 15% solution of n-undecane in petroleum ether (bp 40-60°C) by tank development. The hydrotropes were developed with water saturated with undecane and visualized by spraying with a 1:1 solution of 5 N sulfuric acid and 0.1 N potassium permanganate which, after heating, gave brown spots on an off-white background.

It was found that none of the anionic, cationic, and nonionic surfactants likely to be found in detergent formulations interfered with the separation, since these materials did not migrate from the starting point. Likewise, many inorganic compounds migrated with the solvent front. Certain possible interfering materials did migrate, but the majority did not cause problems because either they were not visualized when present at the concentrations normally found in formulations, or their R_f values were different from those of the hydrotropes.

e. <u>Chromatography of Phosphates</u>. The separation of some 13 phosphate-type surfactants has been described by Kinoshita and Oyama [231] using plates of silica gel modified with disodium ethylenediaminetetraacetic acid. Various solvent combinations were used to develop the samples by ascending chromatography, including chloroform:methanol:acetic acid:water (68:18:11:3), chloroform:methanol:acetic acid:1 N sulfuric acid (74:10:15:1), chloroform:methanol:water:1 N sulfuric acid (72:25:2.5:0.5), chloroform: methanol (9:1), and chloroform:methanol:0.1 N sulfuric acid (96:3.8:0.2).

Good separation without tailing was claimed for alkylphosphates, ethoxylated alkyl ether phosphates and ethoxylated alkylphenol phosphates.

Mention has already been made above (subsection a) of the separation of sulfates, phosphates, and phosphonates on ammonium sulfate modified silica [198].

4. Electrophoresis

Electrophoresis has found little application in the analysis of anionic surfactants compared with paper and thin-layer chromatography, perhaps because the latter techniques require relatively inexpensive equipment and method development is more straightforward.

The first study of electrophoresis applied to anionic surfactants appears to be that by Gasparic et al. [232] who studied the scope of paper electrophoresis for the separation and identification of surfactants in commercial disperse and vat dyes. Paper and thin-layer methods could not be applied to these materials because the dyes were often soluble in the mobile phases used and formed long trailing spots on the chromatogram which overlapped the surfactants under investigation.

For the separation of anionic, cationic, and nonionic surfactants, 2 N acetic acid and 2 N ammonia were found to be the best electrolytes. At a potential of 300 V, separation could be achieved in 2 hr on a 30-cm strip of Whatman No. 3 paper. After drying, anionic surfactants were visualized by spraying with pinacryptol yellow solution. In both electrolytes, the anionics migrated toward the anode while the cationics migrated toward the cathode; the nonionics effectively did not migrate. Separation of all three classes from one another was thus easily achieved.

With regard to the anionics, the migration rate was found to be dependent upon the charge and size of the molecule. Migration increased with increasing numbers of sulfo groups in the molecule but with decreasing length of the aliphatic chain. The separation of homologous series was poor, however, and only effective when significant differences in chain length were present.

Successful separation of the homologous series of alkyl sulfates, alkylbenzenesulfonates, alkyl ether sulfates, and alkylaryl ether sulfates was achieved by Bodenmiller [233, 234] using agarose gel electrophoresis.

Preliminary studies with aqueous gels showed that separation of the various homologous series could not be obtained in this medium because of the existence of monomer-micelle equilibria.

The presence of micelles during migration was eliminated by using agarose gels containing micelle denaturizing organic solvents, including dimethylformamide, n-propanol, p-dioxane, acetamide, cellosolve, ethylene glycol, glycerol, 1,2-propanediol, and urea. From this study it was concluded that gels containing dimethylformamide or dioxane and 1.2% agarose gave the best results. Separation of the members of the homologous series was accomplished using 50% aqueous dioxane, 0.01 M phosphate at pH 6, and 1.2% agarose gels with an applied potential of 1540 V. Migration distances of 25 to 30 cm and periods of 2.0-2.75 hr were required to obtain complete resolution of adjacent members with ionic weights not greater than 325. After separation, the components were detected by precipitation with pinacryptol yellow and identified by comparing their migration distances with those of standards run at the same time. Permanent records of separation profiles could be obtained by photometric scanning of the fluorescent pinacryptol yellow-surfactant adducts.

Such an analysis of homologous series has advantages over paper and thin-layer chromatographic methods. Run times are generally shorter and mobility is strictly dependent on ionic weight.   The sample can also be extracted from the gel for quantitative work.

5. Column Chromatography

Historically, adsorption chromatography on columns is the oldest of the chromatographic methods, dating back to the work of Tswett [235, 236] in 1906. However, in the years following, the technique lapsed into obscurity and little further interest was shown until the late 1930s and early 1940s when the foundations of partition, gas, and thin-layer chromatography were laid.

The nature of adsorbent materials available for column chromatography (e.g., carbon) and the physical size of the columns used (typically 1-4 cm diameter and 10-80 cm long) prolonged analysis, often taking days to effect a separation. Because thin-layer and paper chromatography were much faster and gas chromatography also more sensitive, these techniques received more attention as analytical methods. Column chromatography was mainly used as a preparative technique.

Over the past ten years, column chromatography has experienced a renaissance. The application of the theoretical principles derived from gas chromatography and data gathered from thin-layer chromatography, together with suitable equipment to exploit the technique, have now made liquid chromatography equivalent to gas chromatography in terms of versatility, simplicity, and speed. High-performance liquid chromatography (HPLC), as this "new" technique has become known, has already had many varied applications [237,238] but so far little has been published regarding its capabilities in the analysis of anionic surfactants. It is the belief of the authors that this state of affairs will soon be rectified.

Liquid chromatography (classical and modern) can be divided into four branches:

liquid/solid or adsorption chromatography

liquid/liquid or partition chromatography

ion exchange chromatography

exclusion chromatography

Ion exchange, because of its very wide use in surfactant analysis has been dealt with separately in Sec. III.C.1 and will not be further considered here.

a. Liquid/Solid (Adsorption) Chromatography. This technique can be thought of as the "closed" version of thin-layer chromatography in which components adsorbed onto a powdered solid are displaced by a mobile liquid phase.

In some early applications of adsorption chromatography, Koch [239] and Brooks et al. [87] used columns of Attapulgus clay (an hydrated magnesium

aluminum silicate) in the analysis of oil-soluble petroleum sulfonates.
These materials were adsorbed selectively from petroleum ether [239] or
chloroform [87], but could be completely eluted using methanol [239].

Columns of activated carbon have been used in the determination of
trace anionic surfactants in water. Sallee et al. [68] were able to concentrate alkylbenzenesulfonates (ABS) by this technique which also removed
many interferences in the determination step. The surfactant was desorbed
with an alkaline (0.04 N KOH) solution of benzene:methanol (1:1) for determination by infrared spectrometry. Harker et al. [240] used charcoal
mixed with the ion exchange resin Zeocarb 225 to remove all organic matter
from waste waters. Subsequent elution with the appropriate solvents gave
the desired components. Thus sodium cetyl sulfate was removed with
ethylene glycol and soap fatty acids with ethanol.

Activated carbon for chromatography can vary greatly in carbon content, depending on its source, e.g., sugar, animal, wood, etc. Although
carbon is the main adsorbent in all these materials, its properties are
modified by the presence of other constituents and hence its chromatographic
behavior will vary. In addition, properties vary according to structure
which is dependent on the thermal treatment received during formation.
Care must be taken, therefore, to select a material which exhibits the desired properties and that these are reproducible from batch to batch.

The behavior of fatty acids on charcoal (particularly the wood charcoal
Darco G-60) has been studied extensively by Cassidy [241-243], Hagdahl
[244], Holman [245], and Cason [246]. Whereas Cassidy and Cason effected
separation by solvent elution techniques, Holman and co-workers developed
the techniques of displacement and carrier displacement to remove adsorbed
acids. In displacement chromatography the displacer chosen is a substance
which is more strongly adsorbed than any component of the sample and is
added continuously with the solvent (e.g., palmitic acid, picric acid). The
sample components then emerge from the column as a sequence of bands.
In carrier displacement chromatography, individual displacers are added
to the sample, one for each component, so that the bands of added and
original compounds emerge alternately from the column. Fatty acid esters
have been used to displace fatty acids.

The use of adsorption chromatography for the analysis of fatty acids
has been reviewed by Holman [247], Lederer [248], Schlenk [249], and
Markley [174].

Although silica gel (silicic acid) is perhaps the most widely used of
all chromatographic adsorbents, it has found little application in the analysis
of anionic surfactants since these materials are quite strongly adsorbed and
are therefore difficult to elute. During their investigations into the lower
acids in sewage and waters, however, Mueller et al. [250] found that silicic
acid modified with 0.5 N sulfuric acid also retained anionic surfactants
(alkyl sulfates and alkylarylsulfonates) and that these could be eluted with
15% butanol in chloroform.

The use of silicic acid impregnated with silver nitrate has become very popular for the separation of fatty acids or their methyl esters according to their degree of unsaturation (although TLC is generally preferred for analytical separations). Silicic acid is the most common adsorbent used for this technique, but Florisil, alumina, and ion exchange resins have also been used. The theory and applications of the method have been reviewed comprehensively by both Morris [251] and Guha and Janak [252].

It would appear that many of the problems associated with carbon and silica gel as adsorbents may be overcome by using a cross-linked polystyrene resin, such as Amberlite XAD-2. This is designed to adsorb selectively water-soluble organic compounds by attraction of their hydrophobic portion to the surface of the porous resin. Hydrophilic substances have little affinity for the resin and are not adsorbed. The adsorbed water-soluble organics may be eluted with a polar solvent such as methanol.

The procedure was first utilized by Scoggins and Miller [253] in the analysis of sulfonic acids from synthesis processes. The sample was dissolved in water and an aliquot introduced onto the column. Elution with additional water caused disulfonic acid to pass out of the column, whereas subsequent elution with methanol displaced the monosulfonic acid. The method was applied to alkyl (dodecane) and aryl (methylnaphthalene, naphthalene, and biphenyl) sulfonic acids.

Pollerberg [254] later extended the method to the isolation of surfactants from mixtures with both inorganic and organic substances. Examples of the applications include separation of alkylbenzenesulfonate from toluenesulfonate; determination of surfactants in the presence of inorganic salts, alkanolamine salts, and urea and the separation of soap from alkylbenzenesulfonate. By introducing gradient elution methods, he was able to refine the separations.

These examples of the use of polystyrene column chromatography indicate the possibilities in surfactant analysis and the technique seems worthy of further exploitation.

b. Liquid/Liquid (Partition) Chromatography.

(1) Normal-Phase Partition. In partition chromatography, use is made of the distribution equilibria of solutes between two (immiscible) liquid phases and it thereby resembles the technique of counter-current distribution. However, for chromatography, one liquid phase is held immobile by a solid support while the second is allowed to flow past the immobile phase. If the stationary phase is more polar than the mobile phase, the technique is referred to as normal-phase chromatography.

In practice, the mobile and immobile phases are never completely immiscible and it is necessary to saturate the mobile phase to prevent stripping during chromatography. This can introduce a new problem, namely contamination of sample components with immobile phase. Since the immobile phase usually must be similar in functionality to the sample components to achieve retention, it may be very difficult to remove it from the components of interest when these are required for further determination or analysis. The problem can be overcome either by judicious choice of the immobile phase so that later removal is simplified, although this may limit the choice of phases available, or by using the chemically bonded phases developed more recently for HPLC [237, 238]. With these materials the immobile phase is chemically bonded to the support material and, providing certain solvents are avoided (namely those containing high concentrations of alcohols or water), no presaturation is required and no column bleed is experienced.

In the majority of reported applications to anionic surfactants, classical partition chromatography has been used and solvents have been chosen to produce immobile phases in situ.

Thus Mutter [255] separated active detergent from nonactive material by using the cross-linked dextran Sephadex G-10. In mixed organic solvent/water systems, this material has the property of retaining the aqueous phase in the gel, thereby creating a greater concentration of organic solvent in the mobile phase, exactly the conditions required for a partition system [256].

The sample (sodium alkanemonosulfonate or sodium alkyl sulfate) was freed from nondetergent organic matter by petroleum ether extraction before introduction onto the column. Chromatography was then performed by passing 300 ml of eluent (chloroform:propanol, 1:1 v/v, saturated with water) through the column. The eluate contained only the active detergent which was subsequently determined by titration after conversion to the acid form by means of a second column in series containing the cation exchange resin Biorad AG 50W-X8. Ionic nondetergent components (sodium sulfate, sodium toluenesulfonate, sodium alkanepolysulfonate) remaining on the Sephadex column were removed by elution with 200 ml of ethanol:water (3:5 v/v) and again determined by titration after conversion to the acids. Analysis of standard mixtures showed the method to be highly accurate.

A Sephadex modified by hydroxypropylation (Sephadex LH-20) is also available which, because it has both hydrophilic and lipophilic properties, swells in polar organic solvents as well as in water [257]. In solvent systems containing both polar and nonpolar organic solvents the gel retains mainly the more polar component. Although this material would seem to offer useful properties for detergent analysis by partition, no data relating to anionic detergents have been published.

Sodium alkanemonosulfonates have also been analyzed by partition chromatography on cellulose containing 10% w/w of water [258]. After removal of water and sodium sulfate from the sample by extraction with isopropanol [84] or butanol the sulfonate was converted to the sulfonic acid for chromatography.

Elution of the sample with 180 ml of 5% n-butanol in petroleum ether removed the monosulfonic acids. Di- and polysulfonic acids were then eluted with 160 ml of water. The method is applicable to both primary and secondary C_{10}-C_{20} sulfonates.

At least 5% by wt of water was required to give effective separation, since on dry cellulose the disulfonic acids are weakly adsorbed and are partially eluted with the monosulfonic acids.

In one of the earliest applications of HPLC, Bombaugh and Little [259] describe separations of anionic surfactants on several partition chromatographic systems. Thus sodium methylnaphthalenesulfonate was separated from ammonium nitrate using Sephadex G-25 as the immobile phase and ethanol:water (95:5) as the mobile phase. Columns containing water on Chromosorb, with n-heptane as mobile phase, were useful for the general separation of surfactants. Thus separation of anionic types as well as separation of anionics from nonionic and cationic materials could be obtained, the elution being in the order of their hydrophilic-lipophilic balance index. For the analysis of carboxylic acids and their salts, polyethylene glycol 400 on Celite with n-heptane as mobile phase allowed elution of these materials in order of increasing ionization constant.

A more popular stationary phase for the analysis of carboxylic acids by partition chromatography is modified silica gel. Ramsey and Patterson [260] were among the first to separate the individual fatty acids from their mixture by using a column of silicic acid with furfuryl alcohol and 2-amino-pyridine as immobile phase and hexane as the mobile phase. One disadvantage was that the method did not give complete separation of odd- and even-numbered acids.

This method has been extended and improved by Nijkamp [261, 262] and Zbinovsky [263]. Nijkamp obtained quantitative separation of the saturated fatty acids from C_{10}-C_{24} using silica impregnated with aqueous methanol. The acids were eluted in order of increasing molecular weight with 2,2,4-trimethylpentane (isooctane) saturated with 95% methanol.

Zbinovsky, using columns of silica with ethylene glycol monomethyl ether (methyl cellosolve):water (9:1) as immobile phase and petroleum ether (Skellysolve B), n-butyl ether, or mixtures of the two as mobile phase, obtained quantitative separation of all the saturated fatty acids from C_2 to C_{16} (together with all the dicarboxylic acids from C_2 to C_{22}). For the fatty acids in the range C_{10} to C_{16}, the most suitable solvent was Skellysolve B. In this system the acids are eluted in order of decreasing molecular weight.

More comprehensive reviews of the use of partition chromatography for fatty acid analysis have been given by Holman [247] and Markley [174].

(2) Reversed-Phase Partition. In this form of partition chromatography, the immobile phase is less polar than the mobile phase, and the technique was developed to overcome a fundamental difficulty in the normal-phase partition analysis of fatty acids. The distribution of long-chain fatty acids between polar and nonpolar phases greatly favors the nonpolar phase because of the large hydrocarbon chain present in the acids. Hence separations by normal-phase partition are much less feasible since the acids remain in the less polar mobile phase instead of undergoing efficient distribution between the phases. The changeover point between normal- and reversed-phase systems is at a chain length of 10 to 12 carbon atoms [174].

The technique was developed at about the same time by Boldingh [264] and Howard and Martin [265]. Boldingh used columns of vulcanized rubber (Hevea) impregnated with benzene as the immobile phase and obtained good separation of C_6-C_{18} saturated acids using mixtures of methanol/acetone: water as mobile solvents. Increasing the content of methanol/acetone in the solvent eluted successively longer-chain acids. Problems with the original method caused later workers [266, 267] to modify the mobile phase to acetone:water mixtures.

Howard and Martin used systems of paraffin oil and acetone:water or n-octane and methanol:water. The first system was preferred since it gave a very stable column. Whereas columns of rubber act as their own support, it was necessary to find a hydrophobic support for the paraffin-oil immobile phase. The diatomaceous earth kieselguhr was found to be highly suitable for this purpose, especially after treatment with dichlorodimethylsilane rendered it nonwetting to polar solvents.

Again, increasing the concentration of acetone in the mobile phase accelerated elution of the longer-chain acids. The method was applied to the analysis of both saturated and unsaturated acids.

There have been a number of extensions and modifications to the procedure of Howard and Martin since it is perhaps simpler to perform than Boldingh's method. Popjak and Tietz [268] extended the analysis to C_{10}-C_{18} acids whereas Crombie et al. separated C_8-C_{20} acids [269] and investigated the behavior of unsaturated acids [270]; Silk and Hahn [271] separated the even-numbered acids from C_{16} to C_{24}. Kapitel [272] separated quantitatively all the acids from C_6 to C_{22}, and Badami [273] carried out quantitative analyses using acetylated castor oil as an immobile phase. Privett and Nickell [274] used heptane and acetonitrile:methanol (85:15) as the phases for separation of the acids as their methyl esters.

Because of general dissatisfaction with reproducibility of data obtained from the Howard and Martin system, Vandenheuvel and Vatcher [275], after successive modifications, found silane-treated silicic acid with 2,2,4-trimethylpentane as stationary phase and aqueous methanol as eluting solvent to be much more satisfactory for the analysis of C_{12}-C_{24} fatty acids. They selected a well-defined granular fraction (170-230 U.S. standard mesh) from a commercial silicic acid for their column, and evolved a standard silanizing treatment. Column performance was improved by incorporating 6% by wt of water in the silicic acid before silanation. Since this water reacts with excess dichlorodimethylsilane to form a silicone, it would seem that the actual immobile phase must be a mixture of isooctane plus silicone; silicones are known to perform favorably in reversed-phase columns [249].

By including an automatic gradient elution device in their equipment, Vandenheuvel and Vatcher were able to change the solvent composition continuously from 75 to 100% methanol and obtain quantitative separation of the seven fatty acids under examination in about 5 hr with an error of less than 2.5%.

Normally, it was necessary with all these reversed-phase methods to presaturate the phases outside the column before chromatography to avoid stripping of the immobile phase during elution. To further prevent changes in solubility, the temperature was generally held constant. Temperatures above ambient offered the advantage of increasing the solubility of the fatty acids, thereby improving chromatography [265, 267, 272, 273]. Where the temperature was not strictly controlled (see for example, Ref. 270) some ageing effects were seen after repeated use.

Schlenk and Gellerman [249] found these precautions were less important when using silicone oil as the immobile phase. The oil was held by untreated diatomaceous earth, and acetonitrile:water mixtures varying from 65:35 to 85:15 were used to separate the esters of C_{12}-C_{20} acids. Silanizing the diatomaceous earth, equilibrating the mobile phase with silicone, and thermocontrolling the column offered little advantage.

A novel means of overcoming the need to presaturate phases was devised by Green et al. [276]. They used a column packed with powdered polyethylene (200 mesh B.S.) which acted as a self-supporting reversed immobile phase. Excellent quantitative separations of C_6-C_{20} acids were obtained using aqueous acetone solvents at room temperature for C_6-C_9 acids, and at 35°C for C_{10}-C_{20} acids. To obtain reasonable flow rates it was necessary to apply pressure from a nitrogen cylinder to the head of the column. Because there was no change in the condition of the column during chromatography, the same column could be used many times without deterioration.

In a more recent version of reversed-phase chromatography of fatty acids, hydrophobic Sephadex was used to support the immobile phase [277]. This material is the hydroxyalkoxypropyl ether of Sephadex and is derived from Sephadex LH-20. In mixtures of polar and nonpolar organic solvents, the hydrophobic Sephadex preferentially retains the nonpolar solvent, creating a reversed-phase system. Beijer and Nyström [277] originally tried the familiar aqueous acetone or aqueous methanol mobile phases, but these were unsatisfactory since they did not contain any component with a high affinity for the gel matrix. Solvent systems consisting of water, methanol, and ethylene chloride were found to give much more efficient separations. The system water:methanol:ethylene chloride (30:70:10) was useful for the separation of free fatty acids whereas a 20:80:10 mixture was preferred for fatty acid methyl esters.

Using these solvents, acids and esters in the C_8-C_{18} range could be completely separated, and elution volumes were highly reproducible if a constant temperature was maintained.

In common with both reversed-phase paper and thin-layer chromatography, reversed-phase column chromatography suffers from the problem of "critical pairs," i.e., the incomplete resolution of certain pairs of saturated and unsaturated acids (see Secs. III.C.2.b and III.C.3.b). Their separation on columns can generally be approached in much the same way as in paper and thin-layer chromatography, that is, to chromatograph the original mixture, then derivatize the unsaturated acids and rechromatograph. The original unsaturation in the acids can then be deduced from the change in elution pattern. Derivatization methods used include hydrogenation [268, 271-273, 275], oxygenation, either partial to the point of hydroxylation [265, 267, 273] or by fragmentation at the double bonds [269, 270, 275], and reaction with mercuric acetate [278-280].

Detailed discussions of the applications of reversed-phase column chromatography to fatty acids are given by Markley [174], Badami [281], and Schlenk and Gellerman [249]. Chobanov et al. [282] discussed the choice of solvents for such systems.

Despite the very wide use of this form of chromatography to fatty acid analysis, it has had little application to other anionic surfactant types. The only published work appears to be that of Puschmann [283] on the analysis of olefinsulfonates using silanized silica gel.

The reaction products resulting from the sulfoxidation of straight-chain paraffin mixtures are alkanemonosulfonate, alkanedisulfonate, sodium sulfate, and water. By means of chromatography on silanized silica gel, a technical product could be fractionated into mono- and disulfonates using isopropanol:water (3:7) as mobile phase.

Sulfonation of α-olefins produces more complex reaction products, including alkenesulfonates, hydroxyalkanesulfonates, sulfatosulfonates, and sodium sulfate. Again the same chromatographic system could be used to separate the mono-, di- and polysulfonates. Thin-layer chromatography of fractions collected from the column, on plates of silica modified with ammonium sulfate and with chloroform:methanol:0.1 N sulfuric acid (80:19:1) as developing solvent, showed that the order of elution was sodium sulfate, di- plus poly- plus sulfatosulfonate, and then monosulfonate. For quantitative analysis, bulk fractions of column eluate were collected so that sodium sulfate, di- and polysulfonates were separated from the monosulfonates. The sulfonate contents were then determined gravimetrically and the sodium sulfate by titration.

The reason for the lack of applications of reversed-phase column chromatography to anionic surfactants other than fatty acids is obvious, if one looks at the development of analytical methods over the last 25 years or so. Up to about 1950, most chromatographic analysis was carried out on columns, either by adsorption or normal-partition mechanisms. In 1950, Boldingh, and Howard and Martin, introduced reversed-phase column chromatography which found immediate and extensive use in the analysis of fatty acids. However, the one basic disadvantage of all the column methods was their inherent slowness; separations often lasted for hours and sometimes days. Also at this time paper and thin-layer chromatography were becoming highly developed and showed great potential in terms of quality and speed of analysis. Thus it was only natural for workers interested in surfactant analysis to turn their attention to these methods. Consequently, many methods involving reversed-phase paper and thin-layer chromatography were published (see Secs. III.C.2 and III.C.3) with a subsequent decline in column chromatography.

Today, the development of specialized bonded-phase packings for HPLC offer much for the future of reversed-phase chromatography in surfactant analysis. Because of their nature [237, 238], presaturation of phases is no longer necessary and the stability of the column is such that reproducible analyses can be assured over very long periods. With both pellicular and microparticulate reversed-phase packings, separations may now be achieved which are superior to any obtained by paper or thin-layer chromatography in terms of both time and resolution of components.

c. Steric Exclusion Chromatography. Steric exclusion is a fairly recent development in chromatographic techniques. The concept originated with the work of Wheaton and co-workers [284-286], who found that nonionic materials of low molecular weight could be separated by elution with water through a column of ion exchange resin both one from another and from ionic compounds. Since then column techniques using cross-linked gels (especially the Sephadex type) have been widely used in separations of large from small molecules in aqueous solutions, also known as gel filtration. However, since Sephadex is hydrophilic, it is limited to aqueous systems and extension to use in organic solvents was only possible when porous polystyrene gels became available. In this form the technique is generally known as "gel permeation chromatography."

The mechanism of these systems is the same and depends on a molecular sizing effect caused by the ability of sample constituents to diffuse into the matrix of the solvent-swollen gel. Molecules larger than the maximum pore size cannot penetrate the gel particles and hence pass through the column and are eluted first. Smaller molecules penetrate the gel pores to various degrees, depending on their size and shape. These molecules are therefore eluted in order of decreasing molecular weight.

The application of steric exclusion chromatography to the analysis of anionic surfactants has only found limited attention, probably because most applications have been performed in aqueous systems requiring the use of "soft" Sephadex-type gels, which are limited to low column-inlet pressures because of their compressibility, leading to lengthy procedures. Now a number of rigid hydrophilic gels and controlled-pore-size glasses are available which can withstand much higher pressures, and these should allow high performance chromatography in aqueous solutions.

One of the earliest applications of steric exclusion chromatography is that by Voogt [133] who employed the principle to separate anionic surfactant from the acetate eluent used during ion exchange separation from nonionics and soaps (Sec. III.C.1.a).

For acid-hydrolyzable detergents, Voogt preferred to use sodium or ammonium acetate as eluent in place of hydrochloric acid to prevent decomposition, but was faced with the problem of separating the acetate and detergent. By converting the eluate components to the corresponding acids (using a cation exchange resin) separation was achieved by ion exclusion on a column of Dowex AG 50W-X8(H^+) cation exchange resin. In this system, the acetic acid behaves as a nonelectrolyte because its ionization is suppressed by the high hydrogen ion concentration from the strong acid cation exchanger. Subsequently, on elution with water, the strong anionic surfactant acid is removed from the column first before the acetic acid. The method was applied to eluates containing dodecyl sulfate and myristoyl-methyltaurine with satisfactory results.

In a series of papers, Nakagawa and Jizomoto [287-290] have investigated the gel filtration of anionic surfactants from both a theoretical and an experimental viewpoint. They took the case of a surfactant solution with a concentration high enough to produce micelles. These micelles correspond to large molecules in the filtration process, whereas monomolecularly dispersed molecules that coexist with them correspond to the small molecules. The system was complicated by the disruption of micelles into monomers and the combination of monomers into micelles which precluded obtaining the theoretical elution curve by conventional mathematical techniques. Nakagawa and Jizomoto therefore used a computer to develop a simulation of the gel column and to derive the theoretical elution curve.

For anionic surfactants, elution curves were obtained for the following systems:

1. Anionic surfactant (sodium decyl sulfate) as a single component [287]

2. Anionic surfactant (sodium dodecyl sulfate) in the presence of an added electrolyte (sodium chloride) [288]

3. A mixture of two anionic surfactants (sodium decyl and dodecyl sulfates) [289]

4. A mixture of anionic surfactant (sodium decyl sulfate) with a nonionic surfactant (octyl-β-D-glucoside) [290]

The experimental elution curves were obtained on columns of Sephadex G-25 (fine) by collecting fractions of effluent and measuring the components gravimetrically and by titration. In the first three systems the agreement between theoretical and experimental curves was very good, apart from small differences due to factors such as the flow of solution in the gel pores and adsorption effects which occur in practice but had been neglected in the idealized theoretical conditions. In the last case, however, the agreement was rather unsatisfactory and this was thought to be due to the neglect of complex mixed micelle effects in the theoretical calculation.

In a more practical application of gel filtration, Garvey and Tadros [291] fractionated the condensation products of sodium naphthalene-2-sulfonate and formaldehyde (molecular-weight range 200 to 2000) into narrow molecular-weight cuts on a preparative scale for use as dispersing agents.

Chromatography was performed on Sephadex G-25 (medium) and the elution curve monitored by measuring the absorption at the maximum between 300 and 340 nm for collected fractions. These fractions were further examined by means of TLC and viscosity and molecular-weight determinations.

The separation obtained was rather unusual. Separation in order of decreasing molecular weight should have been obtained but, in fact, the order was found to be nonamer, octamer (heptamer), sodium sulfate, monomer, monomer, dimer to hexamer (heptamer). The reason for this unusual order was explained by the presence of a strong negative charge on the gel due to dissociated acid group impurities and orientated adsorbed water dipoles. For nonionic polymers, Sephadex G-25 has a maximum pore size so that compounds up to molecular weight 5000 can enter the gel pores, but because of the negative charge the polyelectrolyte chains are prevented from entry by electrostatic repulsion. The pore size of the gel was thus effectively reduced to exclude molecular weights greater than 1600. Hence the nonamer, octamer, and possibly heptamer eluted first, whereas the monomer to heptamer (and sodium sulfate) entered the gel structure and were adsorbed by means of their aromatic groups. These molecules were then eluted in order of increasing adsorption strength, i.e., sodium sulfate (nonadsorbed), monomer, dimer, trimer, etc.

An example of the analysis of a sodium lignosulfonate has been given by Bombaugh et al. [292]. Chromatography was performed on columns of Aquapak (a cross-linked polystyrene containing hydrophilic groups) of porosities 7×10^5, 1.5×10^4, and 1.0×10^3 nm with water as eluent.

d. Salting-Out Chromatography. Ion exchange resins have proved to be very versatile packings for chromatographic separations and have been used under the following conditions:

1. Separation of ions by elution with a solution of an electrolyte (ion exchange)

2. Separation of an electrolyte from a nonelectrolyte by elution with water (ion exclusion)

3. Separation of two or more nonelectrolytes by elution with water [286]

In applying the latter conditions of chromatography, Sargent and Rieman [293] found that separations of alcohols were greatly improved by using a salt solution (3 M ammonium sulfate) instead of water. This was explained in terms of partition chromatography. The presence of electrolyte in the eluent reduces the solubility of the organic compounds in the aqueous phase with subsequent displacement into the resin phase. Linear increases in salt concentration were found to cause an exponential increase in the retention volume of the alcohols.

This principle was first applied to surfactant analysis by Keily et al. [294] who separated toluene- and xylenesulfonates from commercial

TABLE 6

Separation of Anionic Surfactants by Salting-out Chromatography

Sample	Column	Eluent	Temp., °C	Ref.
Linear (soft) and branched-chain (hard) alkylbenzenesulfonates	Amberlite CG-50[a] (100–200 mesh) 500 × 26 mm	0.5 M ammonium sulfate and 43% methanol 0.45 ml/min	50	295
Condensation products of β-naphthalenesulfonic acid and formaldehyde	CM–Sephadex[a] (C-50) 1960 × 25 mm	0.005 M magnesium chloride 0.1 ml/min	25	296
Hydroxyalkane- and alkenesulfonates in α-olefinsulfonates	Amberlite CG-50 (100–200 mesh) 400 and 740 × 25 mm	0.5 M sodium chloride and 30% isopropanol 0.3 ml/min	35	297
Linear alkylbenzenesulfonates and alkyl sulfates	Amberlite CG-50 (100–200 mesh) 100 × 25 mm	0.5 M sodium chloride and 40% methanol 0.3 ml/min	37	298
Alkyl sulfates and soaps	Amberlite CG-50 (200–400 mesh) 235 × 25 mm	(a) 0.5 M sodium chloride and 30% isopropanol (b) gradient from (a) to 30% isopropanol 0.3 ml/min	40	299
Alkylsulfates, linear and branched-chain alkylbenzenesulfonates, alkanesulfonates, α-olefinsulfonates, di-2-ethylhexylsulfosuccinate, p-toluenesulfonates, and soaps	Amberlite CG-50 (200–400 mesh) 410 × 25 mm	0.2 M sodium chloride and 30% isopropanol 0.7 ml/min	40	300

[a]Weak cation exchangers.

dodecylbenzenesulfonates with the aid of 1 M ammonium sulfate on Dowex 50-X2(NH$_4^+$) resin. After elution of the short-chain sulfonates, the long-chain sulfonates could be removed with water.

Probably the most extensive investigation of the use of this technique has been the work of Fudano and Konishi who have applied it to the separation and determination of mixtures of several anionic surfactants [295-300]. Their eluting solvents were generally mixtures of electrolytes and short-chain alcohols; the latter served to prevent the surfactant from forming micelles and to reduce the extent to which the monomolecular species were adsorbed onto the resin. Details of the analyses performed by these authors are summarized in Table 6.

In a later paper [301], these authors further investigated the low-condensed components in condensates of mixtures of β-naphthalenesulfonic and β-methylnaphthalenesulfonic acids with formaldehyde. The acids were mixed in the proportions 3:1, 1:1, and 1:3 before condensation and condensates of each acid alone with formaldehyde were used as standards. The fractionation of monomer and dimer of the condensates was performed by the salting-out method given in Table 6. Each fraction was then analyzed by HPLC on a 1-m column of a strong anion exchange resin (Zipax SAX) to further separate the components.

These analyses showed that the main components of the monomer and dimer in the condensates were β-naphthalenesulfonic acid (NS) and the dimer NS-NS, regardless of the ratios of the two acids. The conclusion was drawn that the reaction rate of the condensation of the β-methyl acid was greater than that of naphthalenesulfonic acid.

REFERENCES

1. K. Brass and A. Beyrodt, Chemical technical investigations of problems on dyeing and printing. VIII. Wetting agents and their examination. Monatschr. Textil Ind. Trade Issue II 50, 53-56 (1935).
2. J. Balthazar, Qualitative and quantitative analysis of synthetic detergents. Ing. Chemiste 32, 169-196 (1951); 33, 3-16 (1951).
3. A. Hintermaier, Terminology and classification of surface-active substances. Fette, Seifen, Anstrichmittel 59, 369 (1957).
4. J.-P. Sisley, Index des Huiles Sulfonées et Détergents Moderne, Vol. III, Editions Teintex, Paris, 1960, pp. 105-108.
5. M. J. Rosen and H. A. Goldsmith, Systematic Analysis of Surface-Active Agents, 1st ed., Interscience Publishers, New York, 1960.
6. M. J. Rosen and H. A. Goldsmith, Systematic Analysis of Surface-Active Agents, 2nd ed., Wiley-Interscience, New York, 1972.
7. D. Hummel, Identification and Analysis of Surface-Active Agents by Infrared and Chemical Methods, Text Vol., Interscience Publishers, New York, 1962.

8. M. Quaedvlieg, The classification of surface-active agents. Proc. Intern. Congr. Surface Activity, 3rd, Cologne, Vol. 3, Universitäts-druckerei, G.m.b.h., Mainz, 1961, pp. 7-17.

9. M. Quaedvlieg and C. Paquot, Über die Verträglichkeit der Klassifi-zierungen des CID und der IUPAC. Proc. Intern. Congr. Surface Activity, 4th, Brussels, Vol. 1, Gordon and Breach, New York, 1964, pp. 37-46.

10. G. M. Dyson, A New Notation and Enumeration System for Organic Compounds, Longmans, Green, London, 1947.

11. Rules for IUPAC Notation for Organic Compounds, Longmans, Green, London, 1961.

12. Classification of surface-active compounds. A proposal by the Comité International de la Détergence. Melliand Textilber. 42, 1054-1058, 1192-1193 (1961).

13. W. Langmann and H. Hofmann, Ziffernklassifikation von Tensiden. Tenside 11, 185-194 (1974).

14. M. Keller and L. Frossard, Simplified classification of surface-active agents. Proc. Intern. Congr. Surface Activity, 5th, Barcelona, Vol. 1, Ediciones Unidas S.A., Barcelona, 1968, pp. 27-34.

15. McCutcheon's Detergents and Emulsifiers, North American Edition and International Edition, McCutcheon's Division, Allured Publishing Cor-poration, Glen Rock, New Jersey, published annually.

16. N. W. Tschoegl, Analysis of water-soluble synthetic soaps. Rev. Pure Appl. Chem. 4, 171-206 (1954).

17. J. H. Jones, General colorimetric method for the determination of small quantities of sulfonated or sulfated surface-active compounds. J. Assoc. Offic. Agr. Chemists 28, 389-409 (1945).

18. G. P. Edwards, W. E. Ewers, and W. W. Mansfield, Determination of sodium cetyl sulfate and its solution in water. Analyst 77, 205-207 (1952).

19. K. Meguro, Interaction between dyes and surfactants. I. Coagulation and dispersion of dyes by surfactants. J. Chem. Soc. Japan, Pure Chem. Sect. 77, 72-76 (1956); C.A. 50, 9826a.

20. T. Kondo, K. Meguro, and H. Nito, Interaction between dyes and sur-factants II. Bull. Chem. Soc. Japan 32, 857-861 (1959); C.A. 54, 14730a.

21. K. Burger, Methods for quantitative micro-determination and trace detection of surface-active compounds. I. Detection and determination of very small amounts of anionics and cationics in aqueous solution. Z. Anal. Chem. 196, 15-21 (1963).

22. R. Wickbold, Determination of alkylarenesulfonates by photometric turbidity titration. Seifen, Öle, Fette, Wachse 85, 415-416 (1959).

23. R. Bennewitz and K. Fiedler, Zur Bestimmung von Anion and Kation Tensiden durch Flockungs-Antagonist Titration. Tenside 2, 337-340 (1965).

24. W. Kling and F. Pueschel, Determination of alkyl alcohol sulfates in dilute solution. Melliand Textilber. 15, 21-23 (1934).
25. T. U. Marron and J. Schifferli, Direct volumetric determination of the organic sulfonate content of synthetic detergents. Ind. Eng. Chem., Anal. Ed. 18, 49-50 (1946).
26. B. Wurzschmitt, Determination of capillary-active substances in washing and cleaning materials. Angew. Chem. 62, 40 (1950).
27. G. Linari, Detection, separation and determination of sodium lauryl sulfate. Pharm. Acta. Helv. 46, 410-412 (1971).
28. Y. Takayama, Determination of anionic and nonionic surface-active substances in aqueous solutions. Kogyo Kagaku Zasshi 60, 872-874 (1957); C.A. 53, 10809h.
29. F. Karush and M. Sonenberg, Long-chain alkyl sulfates—colorimetric determination of dilute solutions. Anal. Chem. 22, 175-177 (1950).
30. G. R. Wallin, Colorimetric method for determining a surface-active agent. Anal. Chem. 22, 616-617 (1950).
31. M. R. J. Salton and A. E. Alexander, Estimation of soaps and ionized detergents. Research (London) 2, 247-248 (1949).
32. H. B. Klevens, Analysis of colloidal electrolytes by dye titration. Anal. Chem. 22, 1141-1144 (1950).
33. T. Yamagishi and F. Yanagisawa, Determination of anionic surface-active agents of the sulfonated type. Tokyo-to Ritsu Eisei Kenkyusho Nenpo 17, 98-100 (1965); C.A. 67, 109932.
34. B. M. Milwidsky and S. Holtzman, Rapid method to test soaps/syndets. Soap Chem. Specialities 42, 83-86, 154, 156-158 (1966).
35. M. Bares, Two-phase titration of soap in detergents. Tenside 6, 312-316 (1969).
36. T. Saito, Y. Ozasa, M. Hayashi, and G. Kondo, Spectrophotometric determination of sodium C_{12} benzenesulfonate using Rhodamine 6G. Osaka Kogyo Gijutsu Shikensho Kiho 17, 90-92 (1966); C.A. 66, 30258v.
37. J. van Steveninck and J. C. Riemersma, Determination of long-chain alkyl sulfates as chloroform-soluble Azure A salts. Anal. Chem. 38, 1250-1251 (1966).
38. O. E. McGuire, F. Kent, L. L. Miller, and G. J. Papenmeier, Field test for the analysis of anionic detergents in well water. J. Am. Water Works Assoc. 54, 665-670 (1962).
39. M. E. Turney and D. W. Cannell, Alkaline methylene blue method for the determination of anionic surfactants and amine oxides in detergents. J. Am. Oil Chemists' Soc. 42, 544-546 (1965).
40. R. Geyer, Simple method for determination of anionic detergents with a graduated colorimeter. Z. Fischerei 11, 791-794 (1964).
41. T. Uno and K. Miyajima, Determination of surface-active agents. III. Volumetric determination of anionic surfactants using neutral red as indicator. Chem. Pharm. Bull. (Tokyo) 16, 467-470 (1962); C.A. 58, 679g.

42. T. Uno, K. Miyajima, and C. Nagao, Determination of surface-active agents. IV. Colorimetric determination of sodium lauryl sulfate with neutral red. Yakagaku Zasshi 82, 1017-1020 (1962); C.A. 58, 679h.

43. A. S. Weatherburn, Determination of ionic type of synthetic surface-active agents. Can. Textile J. 71, 45-46 (1954).

44. T. Barr, J. Oliver, and W. V. Stubbings, The determination of surface-active agents in solution. J. Soc. Chem. Ind. 67, 45-48 (1948).

45. G. R. Lewis and L. K. Herndon, Determination of surface-active agents. Sewage Ind. Wastes 24, 1456-1465 (1952).

46. G. P. Edwards and H. E. Ginn, Determination of synthetic detergents in sewage. Sewage Ind. Wastes 26, 945-953 (1954).

47. R. Bennewitz, Determination of ion types of capillary-active materials by means of color indicators. Fette, Seifen, Anstrichmittel 58, 832-833 (1956).

48. M. Aoki and Y. Iwayama, Determination of ionic surface-active agents with dyes. I. Erythrosine and eosine methods. Yakugaku Zasshi 79, 522-526 (1959); C.A. 53, 16562.

49. I. Stojkovic and H. Kukovec, Determination of anionic-active agent concentrations in solutions. Tekstil 13, 69-78 (1964).

50. M. Aoki and Y. Iwayama, Determination of ionic surface-active agents with dyes. III. Neutral red. Yakugaki Zasshi 80, 1745-1749 (1960); C.A. 55, 9912.

51. H. Holness and W. R. Stone, A systematic scheme of semi-micro qualitative analysis for anionic surface-active agents. Analyst 82, 166-176 (1957).

52. M. L. Corrin and W. D. Harkins, Determination of the critical concentration for micelle formation in solutions of colloidal electrolytes by the spectral change of a dye. J. Am. Chem. Soc. 69, 679-683 (1947).

53. M. Raison, The determination of the c.m.c. of anionic detergents by colorimetric titrations with a colored cationic-pinacyanol chloride. Compt. Rend 235, 1129-1130 (1952).

54. K. Peter, Thymol blue as a reagent for anionic detergents. Fette, Seifen, Anstrichmittel 58, 997-1001 (1954).

55. M. Dolezil, Use of fluorescent indicators for the determination of small amounts of surface-active compounds. I. Determination of organic sulfates and sulfonates. Chem. Listy 50, 1588-1592 (1956).

56. R. Egginger, Analytical procedure for identification of detergents. Fleischwirtschaft 52, 626-627 (1972).

57. E. S. Abramovich, Colorimetric determination of anionics. Tekstil Prom. 20, 41 (1960); C.A. 55, 14946.

58. J. Renault and L. Bigot, Identification of anionic sulfur-containing detergents. Compt. Rend 251, 1515-1516 (1960).

59. C. G. Taylor and B. Fryer, The determination of anionic detergents with iron III chelates; applications to sewage and sewage effluents. Analyst 94, 1106-1116 (1969).

60. J. Courtot-Coupez and A. Le Bihan, Determination of anionic detergents in sea water. Anal. Letters 4, 211-219 (1969); Bull. Soc. Chim. 1, 406-411 (1970).

61. E. W. Fowler and T. W. Steele, Determination of anionic surface-active agents in dilute aqueous solution. Natl. Inst. Met. Repub. S. Afr. Lab. Methods, No. 443, 1968 (3 pp.); C.A. 73, 16619b.

62. G. S. Buchanan and J. C. Griffith, Polarographic estimation of anionic detergents. J. Electroanal. Chem. 5, 204-207 (1963).

63. S. Greenberg, Differentiation test for anionic, nonionic and cationic surfactants. Chemist-Analyst 51, 11-14 (1962).

64. S. A. Miller, B. Bann, and A. P. Ponsford, Extraction of the active agent from detergent mixtures. J. Appl. Chem. (London) 1, 523-524 (1951).

65. Y. Izawa and W. Kimura, Analysis of surface-active agents. VIII. Determination of anionics by the Epton and Weatherburn methods. Yukaguku 9, 69-72 (1960).

66. S. R. Epton, Method of analysis for certain surface-active agents. Nature 160, 795-796 (1947).

67. J-Y. Shang, Removal of alkylbenzenesulfonates (ABS) from water. U.S. Pat. 3,247,103 to Sun Oil Co. (1966); C.A. 64, 19182g.

68. E. M. Sallee and members of the Association of American Soap and Glycerine Producers Analytical Subcommittee. Determination of trace amounts of alkylbenzenesulfonates in water. Anal. Chem. 28, 1822-1826 (1956).

69. J. D. Fairing and F. R. Short, Spectrophotometric determination of alkylbenzenesulfonate detergents in surface waters and sewage. Anal. Chem. 28, 1827-1834 (1956).

70. C. P. Ogden, H. L. Webster, and J. Halliday, Determination of biologically soft and hard alkylbenzenesulfonates in detergents and sewage. Analyst 86, 22-29 (1961).

71. R. D. Swisher, Identification and estimation of LAS in waters and effluents. J. Am. Oil Chemists' Soc. 43, 137-140 (1966).

72. I. Matsumoto, H. Hasegawa, S. Togano, and K. Ishiwata, Infrared absorption spectrometry of sodium alkylbenzenesulfonates. Kogyo Kagaku Zasshi 72, 549-552 (1969); C.A. 71, 23147.

73. D. Berkowitz and R. Bernstein, Analysis of soap-synthetic detergent mixtures in bar form. Ind. Eng. Chem., Anal. Ed. 16, 239-241 (1944).

74. G. F. Longman and J. Hilton, Methods for the Analysis of Non-soapy Detergent (NSD) Products, Society for Analytical Chemistry London, Monograph No. 1, 1961.

75. W. H. Simmons and members of the Analytical Methods Committee, Examination of detergent preparations. Analyst 76, 279-286 (1951).

76. A. Hintermaier, Determination of active ingredients in laundering compounds. Fette, Seifen, Anstrichmittel 52, 689-693 (1950).

77. B. M. Milwidsky, Continuous liquid/liquid extraction. Soap Chem. Specialities 45, 79-80, 84, 86, 88, 118 (1969).

78. L. U. Ross and E. W. Blank, Error in the determination of active ingredient of detergent products. J. Am. Oil Chemists' Soc. 34, 70 (1957).

79. B. Wurzschmitt, Systematic and qualitative determination of substances with capillary activity. Z. Anal. Chem. 130, 105-185 (1950).

80. J. Vizern and L. Guillot, Analysis of cleansing and wetting agents on the basis of sulfonated fatty acid amides. Ann. Chim. Anal. Chim. Appl. 23, 235-237 (1941).

81. L. E. Weeks and J. T. Lewis, Analysis of surfactant mixtures. J. Am. Oil Chemists' Soc. 37, 38-142 (1960).

82. C. Kortland and H. F. Dammers, Qualitative and quantitative analysis of mixtures of surface-active agents with special reference to synthetic detergents. J. Am. Oil Chemists' Soc. 32, 58-64 (1955).

83. F. T. Weiss, A. E. O'Donnell, R. J. Shreve, and E. D. Peters, Comprehensive analysis of sodium alkylarylsulfonate detergents. Anal. Chem. 27, 198-205 (1955).

84. American Society for Testing and Materials, Standards on Soap and Other Materials, Part 22; Standards on Petroleum Products, Part 18, Philadelphia, 1972.

85. H. Gerber and J. Sporleder, Determination of sulfate in (products made by) highly sulfonating organic compounds such as hydrocarbons or fatty acids. Textilber. 20, 212 (1939).

86. B. M. Milwidsky, Laboratory apparatus for liquid/liquid extractions. Chem. Ind. (London) 13, 411-412 (1969).

87. F. Brooks, E. D. Peters, and L. Lykken, Analysis of oil-soluble petroleum sulfonates. Extraction-adsorption method. Ind. Eng. Chem., Anal. Ed. 18, 544-547 (1946).

88. R. House and J. L. Darragh, Analysis of synthetic anionic detergent compositions. Anal. Chem. 26, 1492-1497 (1954).

89. R. Hart, Determination of sulfur in surface-active agents. Ind. Eng. Chem., Anal. Ed. 10, 688-689 (1938); Determination of active ingredients and fatty matter in surface-active agents. Ibid. 11, 33-34 (1939).

90. J. Bergeron, R. Derenemesnil, J. Ripert, and G. Monier, Qualitative analysis of anion-active derivatives. Bull. Mens. ITERG 4, 118-130 (1950).

91. G. V. Shirolkar and J. Venkataraman, Wetting agents in textile processing. VIII. Purification and analysis of commercial wetting agents. J. Ind. Chem. Soc., News Ed. 4, 61-71 (1941).

92. H. Jahn, The analysis of aliphatic alcohol sulfonates. Chem. Ztg. 57, 383-384 (1933).

93. American Oil Chemists Society, Official and Tentative Methods, Chicago, 1970.

94. W. Küpfer, J. Jainz, and H. Kelker, Analysis of alkanesulfonates. Tenside 6, 15-21 (1969).

95. G. Reutenauer, Analysis of sulfonated alkyl aryls. Bull. Mens. ITERG 4, 197-199 (1950).
96. T. Noshiro, Y. Izawa, and W. Kimura, Analysis of surface-active agents. XII. Liquid/liquid extraction for determination of surface-active agents. Kogyo Kagaku Zasshi 66, 1206-1210 (1963); C.A. 61, 2064g.
97. E. W. Blank and R. M. Kelley, The analytical separation of surface-active agents. J. Am. Oil Chemists' Soc. 41, 137-139 (1964).
98. H. F. Kohler, Separation of mono- and dialkyl sulfates. U.S. Pat. 2,380,532 to Standard Oil Co. (1945); C.A. 39, 4888.
99. E. Swindells, Analysis of sulfonated fatty alcohols and similar condensation products. Dyer 73, 120-121 (1935).
100. C. L. Hoffpauir and J. H. Kettering, Determination of small amounts of soaps or fatty acids on cotton materials. Am. Dyestuff Reptr. 35, 264-266 (1946).
101. J. C. Harris and F. R. Short, Detection of commercial sodium dodecylbenzenesulfonate in canned foods. Food Technol. 6, 275-278 (1952).
102. E. Gould, An extractant for microamounts of anionic surfactant bound to large amounts of protein, with subsequent spectrophotometric determination. Anal. Chem. 34, 567-571 (1962).
103. J. Haslam and H. A. Willis, Identification and Analysis of Plastics, Iliffe, London, 1965.
104. R. Lemlich, Adsorptive bubble-separation methods. Foam fractionation and allied techniques. Ind. Eng. Chem. 60, 16-29 (1968).
105. C. A. Brunner and R. Lemlich, Foam fractionation. Standard separator and refluxing columns. Ind. Eng. Chem. Fundamentals 2, 297-300 (1963).
106. I. H. Newson, Foam separation. Principles governing surfactant transfer in a continuous foam column. J. Appl. Chem. 16, 43-49 (1966).
107. S. A. Klein and P. H. McGauhey, Detergent removal by surface stripping. J. Water Pollution Control Federation 35, 100-115 (1963).
108. J. L. Rose and J. F. Sebald, Treatment of waste waters by foam fractionation. Tappi 51, 314-321 (1968).
109. E. Rubin and R. Everett, Jr., Sewage plant effluents contaminant removed by foaming. Ind. Eng. Chem. 55, 44-48 (1963).
110. C. A. Brunner and D. G. Stephan, Foam fractionation. Ind. Eng. Chem. 57, 40-48 (1965).
111. D. G. Stephan, Water renovation; some advanced treatment processes. Civil Eng. 35, 46-49 (1965).
112. R. B. Grieves, S. Kelman, W. R. Obermann, and R. K. Wood, Exploratory studies on batch and continuous foam separation. Can. J. Chem. Eng. 41, 252-257 (1963).
113. R. B. Grieves, C. J. Crandall, and R. K. Wood, Foam separation of ABS and other surfactants. Intern. J. Air Water Poll. 8, 501-513 (1964).

114. H. Kishimoto, Foam separation of surface-active substances. I. Fundamental treatments. Kolloid-Z. 192, 66-101 (1963).

115. V. V. Pushkarev, Separation of surfactants and ions from solutions by foaming. Studies in the U.S.S.R. Adsorptive Bubble Separ. Tech., 299-313 (1972).

116. E. Rubin and J. Jorne, Foam separation of solutions containing two ionic surface-active solutes. Ind. Eng. Chem., Fundamentals 8, 474-482 (1969).

117. E. Rubin and D. Melech, Foam fractionation of solutions containing two surfactants in stripping and reflux columns. Can. J. Chem. Eng. 50, 748-753 (1972).

118. A. Okabe and T. Ishii, Removal of anionic surfactants from waste waters. Ger. Pat. (offen.) 2,154,105 (1972); C.A. 77, 38895.

119. R. M. Skomoroski, Separation of surface-active compounds by foam fractionation. J. Chem. Ed. 40, 470-471 (1963).

120. F. Sebba, Ion Flotation, Elsevier, New York, 1962.

121. R. Wickbold, Concentration and separation of surfactants from surface waters by transport in the gas water interface. Tenside 8, 61-63 (1971).

122. H. S. Tomlinson and F. Sebba, Determination of surfactant ions by flotation. Anal. Chim. Acta 27, 596-597 (1962).

123. V. M. Lovell and F. Sebba, Ion flotation method for analysis of some cationic and anionic surfactants below critical micelle concentration. Anal. Chem. 38, 1926-1928 (1966).

124. R. Neu, Analysis of washing and cleansing agents. V. Identification and analysis of laundering raw materials using organic base exchangers. Fette, Seifen, Anstrichmittel 52, 349-352 (1950).

125. S. Takahama and T. Nishida, Analysis of surfactants by the ion-exchange method. A new rapid semimicro method for determination of anionics in the presence of nonionics. Proc. Intern. Congr. Surface Activity, 2nd, London, Vol. 4, Butterworths, London, 1957, pp. 141-147.

126. H. Hempel and H. Kirschnek, The use of ion exchange resins for the separation of surface-active substances with special reference to cationic and nonionic compounds. Fette, Seifen, Anstrichmittel 61, 369-374 (1959).

127. R. Wickbold, Separation of detergent mixtures by means of ion exchange. Seifen, Öle, Fette, Wachse 86, 79-82 (1960).

128. M. J. Rosen, Separation of nonionic surface-active agents from mixtures with anionics by batch ion exchange. Anal. Chem. 29, 1675-1676 (1957).

129. M. J. Rosen, Analysis of mixtures of ionic and nonionic surface-active agents. Separation and recovery of components by batch ion exchange. J. Am. Oil Chemists' Soc. 38, 218-220 (1961).

130. M. E. Ginn and C. L. Church, New columnar and mixed-bed ion exchange methods for surfactant analysis and purification. Anal. Chem. 31, 551-555 (1959).

131. P. Voogt, Application of ion exchangers in detergent analysis. Rec. Trav. Chim. 77, 889-901 (1958).

132. P. Voogt, The use of ion exchangers in detergent analysis II. Rec. Trav. Chim. 78, 899-912 (1959).

133. P. Voogt, The use of ion exchangers in detergent analysis. Proc. Intern. Congr. Surface Activity, 3rd, Cologne, Vol. 3, Universitätsdruckerei, G.m.b.h., Mainz, 1961, pp. 78-88.

134. A. Arpino and V. de Rosa, Analytical chemistry of surfactants. I. Application of ion exchange resins in separating anion-active/nonionic mixtures. Riv. Ital. Sos. Grasse 37, 521-527 (1961).

135. N. Blumer, Quantitative analysis of washing agents with ion exchangers; separation of soaps and anionic and nonionic surface-active agents. Schwiez. Arch. Angew. Wiss. Tech. 29, 171-180 (1963).

136. D. Czaja and F. Awerbuch, Application of ion exchange resins for analysis of compositions of detergents. Tluszcze i Srodki Piorace 8, 31-39 (1964) and 8, 94 (1964); C.A. 62, 2919b.

137. K. Bey, The quantitative analysis of surfactants with the aid of ion exchangers. Fette, Seifen, Anstrichmittel 67, 25-30 (1965).

138. H. König, Separation of mixtures of detergents, especially anion-active detergents. Z. Anal. Chem. 254, 337-345 (1971).

139. J. W. Jenkins, Determination of soaps by ion exchange resins. J. Am. Oil Chemists' Soc. 33, 225-226 (1956).

140. M. Mutter, Analysis of alkane sulfonates by means of ion exchangers. Tenside 5, 138-140 (1968).

141. A. E. O'Donnell, Household detergent analysis. Soap Chem. Specialities 47, 26-28, 51 (1971).

142. S. H. Newburger, The analysis of shampoos. J. Assoc. Offic. Agr. Chemists 41, 664-668 (1958).

143. S. H. Newburger, A Manual of Cosmetic Analysis. Association of Official Agricultural Chemists, Inc., Washington, 1962, pp. 8, 38-43.

144. D. M. Gabriel, Specialized techniques for the analysis of cosmetics and toiletries. J. Soc. Cosmet. Chem. 25, 33-48 (1974).

145. N. Kunimine, Analytical methods of photographic materials. VI. Microanalysis for surface-active agents in photographic emulsions. Nippon Shashin Gakkai Kaishi 27, 177-185 (1964); C.A. 63, 6532h.

146. I. M. Abrams, Removal of anionic surfactant from liquids. U.S. Pat. 3,232,867 to Diamond Alkali Co. (1966); C.A. 64, 12358.

147. W. Hughes, S. Frost, and V. W. Reid, Analysis of alkylbenzenesulfonates present in sewage. Proc. Intern. Congr. Surface Activity, 5th, Barcelona, Vol. 1, Ediciones Unidas S.A., Barcelona, 1968, pp. 317-325.

148. H. N. Dunning, J. M. White, and H. Wittcoff, Polymeric liquid anion exchange process. U.S. Pat. 3,215,625 to General Mills, Inc. (1965); C.A. 64, 19182.

149. J. P. Riley and D. Taylor, Analytical concentration of traces of dissolved organic materials from sea water with Amberlite XAD-1 resin. Anal. Chim. Acta 46, 307-309 (1969).

150. S. Wiktorowski and A. Justat, Ion exchange between anion-active substances and strongly basic ion exchangers. Gaz, Woda Tech. Sanit. 43, 52-55 (1969); C.A. 70, 118449b.

151. S. Wiktorowski and A. Justat, Investigation of ion exchange between anion-active substances and strong alkaline ion exchangers II. Gaz, Woda Tech. Sanit. 43, 240-242 (1969); C.A. 71, 129137v.

152. S. S. Shamanaev, V. V. Pushkarev, and N. N. Pustovalov, Nonexchange sorption of surfactants. Zh. Fiz. Khim. 47, 229-231 (1973); C.A. 78, 161171g.

153. S. S. Shamanaev, V. V. Pushkarev, and N. N. Pustovalov, Sorption of surfactants by organic ion exchangers. Zh. Fiz. Khim. 47, 2049-2051 (1973); C.A. 79, 149778y.

154. S. S. Shamanaev, V. V. Pushkarev, and N. N. Pustovalov, Sorption of surfactants by anion exchanger. Zh. Fiz. Khim. 47, 2340-2343 (1973); C.A. 79, 149802b.

155. J. Drewry, Qualitative examination of detergents by paper chromatography. Analyst 88, 225-231 (1963).

156. J. Drewry, Examination of detergents by paper chromatography, Part II. Analyst 89, 75-76 (1964).

157. A. Kopecky, Paper chromatography of higher fatty acids. Prumysl Potravin 9, 385-386 (1958); Anal. Abstr. 6, No. 1936 (1959).

158. R. Chayen and E. M. Linday, A modified method for the paper chromatography of long-chain fatty acids. J. Chromatogr. 3, 503-504 (1960).

159. R. D. Tiwari and K. C. Srivastava, Visualization reagents in the paper chromatography of saturated fatty acids C_{12} to C_{20}. Fres. Z. Anal. Chem. 230, 361-362 (1967).

160. R. D. Tiwari and K. C. Srivastava, Paper chromatography of fatty acids. Use of acetates of cobalt, copper and nickel and of mercurous nitrate as spot detecting reagents. Fres. Z. Anal. Chem. 232, 117-118 (1967).

161. H. P. Kaufmann and W. H. Nitsch, Paper chromatography in the fat field. XVI. Further experiments on the separation of fatty acids. Fette, Seifen, Anstrichmittel 56, 154-158 (1954).

162. H. P. Kaufmann and A. A. Karabatur, A new method for qualitative and quantitative chromatography of critical fatty-acid partners. Nahrung 2, 61-75 (1958).

163. J. Sliwiok, Chromatographic separation of higher fatty acids. Roczniki Chem. 37, 1497-1501 (1963); C.A. 60, 9879c.

164. H. K. Mangold, B. G. Lamp, and H. Schlenk, Indicators for the paper chromatography of lipids. J. Am. Chem. Soc. 77, 6070-6072 (1955).

165. H. Schlenk, J. L. Gellerman, J. A. Tillotson, and H. K. Mangold, Paper chromatography of lipids. J. Am. Oil Chemists' Soc. 34, 377-386 (1957).

166. F. Franks, Paper chromatography with continuous change in solvent composition. Part 1. Separation of fatty acids. Analyst 81, 384-390 (1956).

167. H. P. Kaufmann and M. Arens, Paper chromatography of fatty acids. XXVIII. Separation of thiocyanogen derivatives. Fette, Seifen, Anstrichmittel 60, 803-806 (1958).

168. W. Awe and B. Grote, Paper chromatography of thiocyanogen derivatives of fatty acids. Fette, Seifen, Anstrichmittel 60, 806-809 (1958).

169. M. A. Buchanan, Paper chromatography of the saturated fatty acids. Anal. Chem. 31, 1616-1618 (1959).

170. H. P. Kaufmann and M. M. Deshpande, Paper chromatography of fats. XXVI. The quantitative paper chromatographic-polarographic analysis of fatty acids. Fette, Seifen, Anstrichmittel 60, 537-541 (1958).

171. A. Seher, Determination of paper chromatographically separated long-chain carboxylic acids by photometric means. Fette, Seifen, Anstrichmittel 58, 498-504 (1956).

172. J. Churacek, F. Kopecny, M. Kulhavy, and M. Jurecek, Paper chromatography of carboxylic acids. V. Paper chromatographic separation and identification of the C_{1-16} fatty acids as N,N-dimethyl-p-aminophenylazophenacyl esters. Z. Anal. Chem. 208, 102-116 (1965).

173. H. P. Kaufmann, Paper chromatography in the fat field. XIX. Quantitative paper chromatographic determination of the straight-chain fatty acids and their mixtures. Fette, Seifen, Anstrichmittel 58, 492-498 (1956).

174. K. S. Markley (ed.), Fatty Acids, Their Chemistry, Properties, Production and Uses, 2nd ed., Wiley-Interscience, New York, 1964, Part 3, Chapter XX, pp. 2125-2247.

175. S. Patai (ed.), The Chemistry of Carboxylic Acids and Esters, Wiley-Interscience, New York, 1969, pp. 899-900.

176. F. D. Gunstone, An Introduction to the Chemistry and Biochemistry of Fatty Acids and their Glycerides, 2nd ed., Chapman and Hall, London, 1967, pp. 31-36.

177. H. Holness and W. R. Stone, Separation within the homologous series of alkyl sulfates, alkyl pyridinium and alkyl trimethylammonium halides by paper chromatography. Nature 176, 604 (1955).

178. F. Franks, Partition chromatography of synthetic detergents. Nature 176, 693-694 (1955).

179. F. Franks, Paper chromatography with continuous change in solvent composition. Part II. Separation of surface-active agents. Analyst 81, 390-393 (1956).

180. J. Borecky, Paper chromatography of sulfated fatty alcohols. Chem. Ind. (London) 265 (1962).

181. J. Borecky, Identification of organic compounds. XLVII. Identification and separation of aliphatic C_{10-18} alcohols as monoalkyl sulfates by paper chromatography. Coll. Czech. Chem. Commun. 27, 2761-2764 (1962).

182. J. Borecky, J. Gasparic, and M. Vecera, Identification of organic compounds. XXV. Identification and separation of aliphatic C_{1-18} alcohols by paper chromatography. Chem. Listy 52, 1283-1288 (1958).

183. J. Gasparic and J. Borecky, Identification of organic compounds. XLI. Paper chromatographic separation and identification of alcohols, glycols, polyethylene glycols, phenols, mercaptans and amines as their 3,5-dinitrobenzoates. J. Chromatogr. 5, 466-499 (1961).

184. J. Borecky, Paper chromatography in the analysis of surface-active agents. Abhandl. Deut. Akad. Wiss. Berlin Kl. Chem., Geol. Biol., 159-169 (1966).

185. J. Borecky, Paper chromatography for the analysis of surface-active agents. Kolor Ert. 8, 386-393 (1966).

186. J. Borecky, The use of paper chromatography for the analysis of surface-active compounds. Tenside 3, 189 (1966).

187. B. Sewell, The identification of Teepol and sodium dodecyl sulfate in detergent mixtures. Lab. Pract. 9, 381 (1960).

188. F. Pueschel and D. Prescher, High molecular weight aliphatic sulfonic acids. VII. Paper chromatography of sulfonates and some alkyl sulfates. J. Chromatogr. 32, 337-345 (1968).

189. J. Borecky, Pinacryptol yellow—a suitable reagent for the detection of aryl sulfonic acids on chromatograms. J. Chromatogr. 2, 612-614 (1959).

190. J. Borecky, Identification of organic compounds. XLIX. Separation and identification of anion active compounds by means of paper chromatography. Coll. Czech. Chem. Commun. 28, 229-240 (1963).

191. J. Borecky, Identification of organic compounds. LIII. Identification of alkylarenesulfonates by paper chromatography of the phenols formed by alkali fusion. Mikrochim. Acta, 1137-1143 (1962).

192. J. Borecky, Identification of organic compounds. XLIX. Identification of alkylphenols in their ethylene oxide condensates. Mikrochim. Acta, 824-829 (1962).

193. J. Borecky, Identification of products from waste cellulose lye. Chem. Prumysl 13, 248-249 (1963).

194. F. Pueschel and O. Todorov, Connection between the constitution and certain properties of surface-active benzenesulfonates with heteroatoms

in the aliphatic side chain. I. Solubility, surface activity, critical micelle formation, concentration and paper chromatography of sulfonates. Tenside 5, 193-198 (1968).

195. C. M. Coyne and G. A. Maw, The paper chromatography of aliphatic sulfonates. J. Chromatogr. 14, 552-555 (1964).

196. I. A. Kuzin, V. N. Romanovskii, A. M. Semushkin, and V. A. Torutev, Determination of bis(2-ethylhexyl) phosphate, mono(2-ethylhexyl) phosphate and ortho-phosphoric acid by paper chromatography. Zh. Analit. Khim. 24, 800-802 (1969); C.A. 71, 56464n.

197. E. Stahl (ed.), Thin Layer Chromatography. A Laboratory Handbook, 2nd ed., George Allen and Unwin, London, 1969.

198. H. K. Mangold and R. Kammereck, New methods of analyzing industrial aliphatic lipids. J. Am. Oil Chemists' Soc. 39, 201-206 (1962).

199. C. T. Desmond and W. T. Borden, Identification of surface-active agents in admixture by thin layer chromatography. J. Am. Oil Chemists Soc. 41, 552-553 (1964).

200. A. Arpino and V. de Rosa, Proc. Intern. Congr. Surface Activity 4th, Brussels, Vol. 1, Gordon and Breach, New York, 1967, pp. 497-505.

201. G. Löser, Separation of surfactant mixtures by means of thin layer chromatography. Seifen, Öle, Fette, Wachse 91, 728-730 (1965).

202. K. Bey, Thin layer chromatographic analysis in the field of surfactants. Fette, Seifen, Anstrichmittel 67, 217-221 (1965).

203. K. Bey, Thin layer chromatographic analysis of surfactants. Tenside 2, 373-375 (1965).

204. L. Daradics, Separating surfactant mixtures by thin layer chromatography. Chim. Anal. (Bucharest) 2, 14-17 (1972); C.A. 77, 77051q.

205. K. Yoshikawa, T. Nishina, and K. Takehana, Thin layer chromatography of detergents. I. Simultaneous analysis of commercial liquid shampoos and dishwashing detergents by TLC. Yukaguku 15, 65-72 (1966); C.A. 64, 11445h.

206. R. Egginger and G. Weidauer, Characterization of textile auxillaries (tensides) and their identification with thin layer chromatography. Spinner, Weber Textilveredl. 87, 1084-1088 (1969).

207. T. Takagi and K. Fukuzumi, Application of thin layer chromatography to oil chemistry. I. Chromatography of synthetic surfactants by silica plates. Yukaguku 13, 520-523 (1964); C.A. 63, 18485f.

208. H. K. Mangold, Thin layer chromatography of lipids. J. Am. Oil Chemists' Soc. 38, 708-727 (1961).

209. H. K. Mangold, Thin layer chromatography of lipids. J. Am. Oil Chemists' Soc. 41, 762-773 (1964).

210. M. C. Williams and R. Reiser, Symposium on the methodology of fats and oils. The chemical and biological assay of essential fatty acids. J. Am. Oil Chemists' Soc. 40, 237-241 (1963).

211. A. T. James, Methods of separation of long-chain unsaturated fatty acids. Analyst 88, 572-582 (1963).

212. B. W. Nichols, The separation of lipids by thin layer chromatography. Lab. Pract. <u>13</u>, 299-305 (1964).
213. M. W. Roomi, M. R. Subbaram, and K. T. Achaya, Separation of fatty acetylenic, ethylenic and saturated compounds by thin layer chromatography. J. Chromatogr. <u>16</u>, 106-110 (1964).
214. E. J. Singh and L. L. Gershbein, Determination of the carbon number of n-fatty acids by TLC. J. Chem. Educ. <u>43</u>, 29 (1966).
215. D. C. Malins and H. K. Mangold, Analysis of complex lipid mixtures by thin layer chromatography and complementary methods. J. Am. Oil Chemists' Soc. <u>37</u>, 576-578 (1960).
216. D. Heusser, Thin layer chromatography of fatty acids on silanized silica gel. J. Chromatogr. <u>33</u>, 62-69 (1968).
217. H. P. Kaufmann and Z. Makus, Separation of lipids by thin layer chromatography. Fette, Seifen, Anstrichmittel <u>62</u>, 1014-1020 (1960).
218. H. P. Kaufmann, Z. Makus, and F. Deicke, Thin layer chromatography of fats. II. Separation of cholesteryl fatty acid esters. Fette, Seifen, Anstrichmittel <u>63</u>, 235-238 (1961).
219. H. P. Kaufmann, Z. Makus, and T. H. Khoe, Thin layer chromatography in the field of fats. III. Visualization of substances being examined on the plate. Fette, Seifen, Anstrichmittel <u>63</u>, 689-691 (1961).
220. H. P. Kaufmann, Z. Makus, and T. H. Khoe, Thin layer chromatography in the field of fats. VI. Hydrogenation and bromination on the plate. Fette, Seifen, Anstrichmittel <u>64</u>, 1-5 (1962).
221. H. P. Kaufmann and T. H. Khoe, Thin layer chromatography in the field of fats. VII. Separation of fatty acids and triglycerides on gypsum plates. Fette, Seifen, Anstrichmittel <u>64</u>, 81-85 (1962).
222. T. W. Hammonds and G. Shone, The separation of fatty acid methyl esters (including "critical pairs") by thin layer partition chromatography. J. Chromatogr. <u>15</u>, 200-203 (1964).
223. H. K. Mangold and R. Kammereck, Separation, identification and quantitative analysis of fatty acids by thin layer chromatography and gas-liquid chromatography. Chem. Ind. (London), 1032-1034 (1961).
224. L. J. Morris, Separation of higher fatty acid isomers and vinylogues by thin layer chromatography. Chem. Ind. (London), 1238-1240 (1962).
225. L. J. Morris, Specific separations by chromatography on impregnated thin layers. Lab. Pract. <u>13</u>, 284-289 (1964).
226. L. D. Bergelson, E. V. Dyatlovitskaya, and V. V. Voronkova, Complete structural analysis of fatty acid mixtures by thin layer chromatography. J. Chromatogr. <u>15</u>, 191-199 (1964).
227. M. M. Paulose, The thin layer chromatographic separation of fatty acid methyl esters according to both chain length and unsaturation. J. Chromatogr. <u>21</u>, 141-143 (1966).

228. A. C. Breyer, M. Fischl, and E. J. Seltzer, A systematic study of the variables involved in the reversed-phase thin layer chromatography of oxyethylated alkyl sulfate surfactants. J. Chromatogr. 82, 37-52 (1973).

229. M. Mutter and K. W. Han, Quantitative thin layer chromatography of organic sulfonates. Chromatographia 2, 172-175 (1969).

230. E. Dunn and P. Robson, Identification of hydrotropes in detergent formulations by reversed-phase thin layer chromatography. J. Chromatogr. 27, 300-302 (1967).

231. S. Kinoshita and M. Oyama, Separation of phosphate-type surfactants on silica gel thin layers pre-treated with di-sodium EDTA. Kogyo Kogaku Zasshi 69, 2022-2023 (1966); C.A. 66, 67054b.

232. J. Gasparic, J. Kadlecova, and J. Borecky, Paper electrophoresis of surface-active agents. Proc. Intern. Congr. Surface Activity, 5th, Barcelona, Vol. 1, Ediciones Unidas S.A. Barcelona, 1968, pp. 305-306.

233. J. R. Bodenmiller and H. W. Latz, Electrophoretic separation of alkyl sulfate, alkylbenzenesulfonate and alkylethoxy-sulfate homologs, using aqueous dioxan agarose gels. Anal. Chem. 43, 1354-1361 (1971).

234. J. R. Bodenmiller, A study of the electrophoretic behavior of anionic surface-active agents and the application of electrophoresis to their qualitative and quantitative analysis. Dissertation Abstr. Intern. (B) 32, 5071 (1972).

235. M. Tswett, Physical chemistry studies on chlorophyll adsorption. Ber. Deut. Botan. Ges. 24, 316 (1906).

236. M. Tswett, Adsorption analysis and chromatographic methods. Application to the chemistry of chlorophyll. Ber. Deut. Botan. Ges. 24, 384 (1906).

237. J. J. Kirkland (ed.), Modern Practice of Liquid Chromatography, Wiley-Interscience, New York, 1971.

238. N. Hadden and F. Baumann (eds.), Basic Liquid Chromatography, Varian Aerograph, 1972.

239. J. M. Koch, Analysis of petroleum oil-soluble sodium sulfonates by adsorption. Ind. Eng. Chem., Anal. Ed. 16, 25-28 (1944).

240. R. P. Harker, J. M. Heaps, and J. L. Horner, Use of adsorption columns in the analysis of soap and detergent stabilized emulsions. Nature 173, 634-635 (1954).

241. H. G. Cassidy, Adsorption analysis. II. Adsorption of higher fatty acids. J. Am. Chem. Soc. 62, 3073-3076 (1940).

242. H. G. Cassidy, Adsorption analysis. III. Relation between adsorption isotherm and position on the adsorption column. J. Am. Chem. Soc. 62, 3076-3079 (1940).

243. H. G. Cassidy, Adsorption analysis. IV. Separation of mixtures of higher saturated fatty acids. J. Am. Chem. Soc. 63, 2735-2739 (1941).

244. L. Hagdahl and R. T. Holman, Displacement analysis of lipids. II. Increased separability of fatty acids by depressed solubility. J. Am. Chem. Soc. 72, 701-705 (1950).

245. R. T. Holman, Displacement analysis of lipids. IV. Carrier displacement of saturated fatty acids. J. Am. Chem. Soc. 73, 1261-1263 (1951).

246. J. Cason and G. A. Gillies, Adsorption and chromatography of fatty acids on charcoal. J. Org. Chem. 20, 419-427 (1955).

247. R. T. Holman, Progress in the Chemistry of Fats and Other Lipids, Vol. 1, Pergamon, New York, 1952.

248. E. Lederer and M. Lederer, Chromatography, American Elsevier, New York, 1954.

249. H. Schlenk and J. L. Gellerman, Column chromatography of fatty acids. J. Am. Oil Chemists' Soc. 38, 555-562 (1961).

250. H. F. Mueller, T. E. Larson, and W. J. Lennarz, Chromatographic identification and determination of organic acids in water. Anal. Chem. 30, 41-44 (1958).

251. L. J. Morris, Separation of lipids by silver ion chromatography. J. Lipid Res. 7, 717-732 (1966).

252. O. K. Guha and J. Janak, Charge-transfer complexes of metals in the chromatographic separation of organic compounds. J. Chromatogr. 68, 325-343 (1972).

253. M. W. Scoggins and J. W. Miller, Rapid separation technique for mono- and disulfonic acids. Anal. Chem. 40, 1155-1157 (1968).

254. J. Pollerberg, Adsorption chromatography with columns of polystyrene in the analysis of tensides. Proc. Intern. Congr. Surface Activity, 5th, Barcelona, Vol. 1, Ediciones Unidas S.A. Barcelona, 1968, pp. 327-332.

255. M. Mutter, Analysis of anionic detergents by liquid chromatography using a Sephadex G-10 column. Chromatographia 2, 208-211 (1969).

256. Pharmacia Monograph. Sephadex. Gel filtration in theory and practice. Pharmacia Fine Chemicals, Uppsala, Sweden, 1974.

257. Pharmacia Monograph. Sephadex LH-20 chromatography in organic solvents. Pharmacia Fine Chemicals, Uppsala, Sweden, 1973.

258. W. R. Ali and P. T. Laurence, Column partition chromatographic determination of sodium alkane monosulfonates. Anal. Chem. 45, 2426-2428 (1973).

259. K. J. Bombaugh and J. N. Little, An investigation of liquid-liquid chromatography with a recording detector. J. Chromatogr. 16, 47-54 (1964).

260. L. L. Ramsey and W. I. Patterson, Separation of the saturated straight-chain fatty acids C_{11} to C_{19}. J. Assoc. Offic. Agr. Chemists 31, 441-452 (1948).

261. H. J. Nijkamp, A simple chromatographic method for the determination of the C_{10} to C_{20} saturated straight-chain fatty acids. Nature 172, 1102-1103 (1953).

262. H. J. Nijkamp, A simple chromatographic method for the determination of the saturated straight-chain fatty acids C_{10} to C_{24}. Anal. Chim. Acta 10, 448-458 (1954).

263. V. Zbinovsky, New solvent system for separating monocarboxylic acids (C_2 to C_{16}) and dicarboxylic acids (C_2 to C_{22}). Anal. Chem. 27, 764-768 (1955).

264. J. Boldingh, Fatty acid analysis by partition chromatography. Rec. Trav. Chim. 69, 247-261 (1950).

265. G. A. Howard and A. J. P. Martin, The separation of the C_{12} to C_{18} fatty acids by reversed-phase partition chromatography. Biochem. J. 46, 532-538 (1950).

266. J. H. Van de Kamer, N. A. Pikaar, A. Bolssens-Frankena, C. Couvée-Ploeg, and L. Van Ginkel, Quantitative determination of the different higher saturated fatty acids in fat from blood, chyle and faeces, by means of partition chromatography on rubber. Biochem. J. 61, 180-186 (1955).

267. K. Hofmann, C-Y. Y. Hsiao, D. B. Henis, and C. Panos, The estimation of the fatty acid composition of bacterial lipides. J. Biol. Chem. 217, 49-60 (1955).

268. G. Popjak and A. Tietz, Biosynthesis of fatty acids by slices and cell-free suspensions of mammary gland. Biochem. J. 56, 46-54 (1954).

269. W. M. L. Crombie, R. Comber, and S. G. Boatman, A chromatographic method for the estimation of oleic and linoleic acids in the presence of straight-chain saturated fatty acids. Nature 174, 181-182 (1954).

270. W. M. L. Crombie, R. Comber, and S. G. Boatman, The estimation of unsaturated fatty acids by reversed-phase partition chromatography. Biochem. J. 59, 309-316 (1955).

271. M. H. Silk and H. H. Hahn, The resolution of mixtures of C_{16} to C_{24} normal-chain fatty acids by reversed-phase partition chromatography. Biochem. J. 56, 406-410 (1954).

272. W. Kapitel, Chromatographic separation of mixtures of fatty acids for analytical purposes. Fette, Seifen, Anstrichmittel 58, 91-94 (1956).

273. R. C. Badami, Reversed-phase partition column chromatography: a method for the quantitative analysis of fatty acids. Chem. Ind. (London), 1920-1921 (1964).

274. O. S. Privett and E. C. Nickell, Preparation of highly purified fatty acids via liquid-liquid partition chromatography. J. Am. Oil Chemists' Soc. 40, 189-193 (1963).

275. F. A. Vandenheuvel and D. R. Vatcher, Partition chromatography of aliphatic acids. Quantitative resolution of normal-chain even acids from C_{12} to C_{24}. Anal. Chem. 28, 838-845 (1956).

276. T. Green, F. O. Howitt, and R. Preston, The use of polythene in the separation of fatty acids by reversed-phase chromatography. Chem. Ind. (London), 591-592 (1955).
277. K. Beijer and E. Nyström, Reversed-phase chromatography of fatty acids on hydrophobic Sephadex. Anal. Biochem. 48, 1-8 (1972).
278. E. Jantzen and H. Andreas, Reaction of unsaturated fatty acids with mercuric acetate. Use for preparative separations. II. Chem. Ber. 94, 628-633 (1961).
279. E. Jantzen, H. Andreas, K. Morgenstern, and W. Roth, The separation of unsaturated fatty acids by means of the mercury adducts of their methyl esters. Fette, Seifen, Anstrichmittel 63, 685-688 (1961).
280. H. B. White and F. W. Quackenbush, Separation of fatty ester-mercuric acetate adducts of alumina. J. Am. Oil Chemists' Soc. 39, 511-513 (1962).
281. R. C. Badami, Reversed-phase partition column chromatography: a review. Chem. Ind. (London), 1211-1213 (1965).
282. D. Chobanov, A. Popov, and E. Chooparova, Choice of solvent system in reversed-phase chromatography: separation of normal saturated monocarboxylic acids. Fette, Seifen, Anstrichmittel 68, 85-91 (1966).
283. H. Puschmann, Analysis of olefinsulfonates. Fette, Seifen, Anstrichmittel 75, 434-437 (1973).
284. R. M. Wheaton and W. C. Bauman, Ion exclusion. Ind. Eng. Chem. 45, 228-233 (1953).
285. R. M. Wheaton and W. C. Bauman, Nonionic separations with ion exchange resins. Ann. N.Y. Acad. Sci. 57, 159-176 (1953).
286. D. W. Simpson and R. M. Wheaton, Ion exclusion-column analysis. Chem. Eng. Progr. 50, 45-49 (1954).
287. T. Nakagawa and H. Jizomoto, Simulation applied to the gel filtration of surfactants. Proc. Intern. Congr. Surface Activity, 5th, Barcelona, Vol. 1, Ediciones Unidas S.A. Barcelona, 1968, pp. 299-304.
288. T. Nakagawa and H. Jizomoto, Simulation applied to the gel filtration of surfactants. Part 3. Kolloid-Z. Z. Polymere 236, 79-83 (1970).
289. T. Nakagawa and H. Jizomoto, Simulation applied to the gel filtration of surfactants. Part 4. Kolloid-Z. Z. Polymere 239, 606-610 (1970).
290. T. Nakagawa and H. Jizomoto, Gel filtration of surfactants. J. Am. Oil Chemists' Soc. 48, 571-577 (1971).
291. M. J. Garvey and T. F. Tadros, Fractionation of the condensates of sodium naphthalene-2-sulfonate and formaldehyde by gel permeation chromatography. Kolloid-Z. Z. Polymere 250, 967-972 (1972).
292. K. J. Bombaugh, W. A. Dark, and R. N. King, Gel permeation chromatography: new applications and techniques. J. Polymer Sci., Polymer Symposium No. 21 on Analytical Gel Permeation Chromatography (1968).

293. R. Sargent and W. Rieman, A new technique for the chromatographic separation of organic compounds. J. Org. Chem. $\underline{21}$, 594-595 (1956).

294. H. J. Keily, A. L. Garcia, and R. N. Peterson, The separation of alkylarylsulfonates by salting-out chromatography. Pittsburgh Conf. Anal. Chem. and Appl. Spectroscopy, 1963, Abstract No. 162, pp. 76-77.

295. S. Fudano and K. Konishi, Separation and determination of linear and branched-chain alkylbenzenesulfonates by salting-out chromatography. J. Chromatogr. $\underline{51}$, 211-218 (1970).

296. S. Fudano and K. Konishi, Separation of the condensation products of β-naphthalenesulfonic acid and formaldehyde by salting-out chromatography. J. Chromatogr. $\underline{56}$, 51-58 (1971).

297. S. Fudano and K. Konishi, Separation and determination of α-olefinsulfonates by salting-out chromatography. J. Chromatogr. $\underline{62}$, 467-470 (1971).

298. S. Fudano and K. Konishi, Separation and determination of linear alkylbenzenesulfonates and alkylsulfates by salting-out chromatography. J. Chromatogr. $\underline{66}$, 153-155 (1972).

299. S. Fudano and K. Konishi, Separation and determination of alkylsulfate and soap by salting-out chromatography. J. Chromatogr. $\underline{71}$, 93-100 (1972).

300. S. Fudano and K. Konishi, Separation and determination of mixtures of anionic surface-active agents by salting-out chromatography. J. Chromatogr. $\underline{77}$, 351-355 (1973).

301. S. Fudano, Y. Miyabata, and K. Konishi, Analysis of low condensed components in condensates of mixtures of β-naphthalenesulfonic and β-methylnaphthalenesulfonic acids with formaldehyde. J. Am. Oil Chemists' Soc. $\underline{51}$, 514-515 (1974).

Chapter 2

IDENTIFICATION BY ABSORPTION
SPECTROSCOPY TECHNIQUES

Delia M. Gabriel and V. John Mulley

Unilever Research
Isleworth Laboratory
Isleworth, Middlesex, United Kingdom

I. IDENTIFICATION BY INFRARED SPECTROSCOPY

The use of infrared spectroscopy for the identification of surfactants has
grown over the years and is now one of the most widely practiced physicochemical techniques. Its main advantages over other methods are the relative ease of sampling and of production and interpretation of spectra, the

latter arising from the ability of functional groups in a molecule to be recognized by their absorption of infrared radiation at specific frequencies. It is also possible in some cases to determine the length of the carbon chain.

An important fact to bear in mind when considering the use of the infrared method is that in the case of a mixture of several components the spectrum becomes complex and diffuse, and it is increasingly difficult to recognize anything other than a particular functional group. With increasing sample purity, a clearer spectrum is obtained which leads to easier identification. Consequently, the infrared spectrum of a commercial surfactant may bear only slight resemblance to its isolated major components. A cautious approach must therefore be adopted in the interpretation of spectra, if significant amounts of other components are likely to be present. This problem is highlighted by the fact that many spectra published in the literature are of natural or commercial products; the former are usually mixtures and the latter are often "purified" by relatively crude separation methods.

A. The Principles of Infrared Absorption Spectroscopy

The interaction of electromagnetic radiation with the natural vibrations of polyatomic molecules produces an absorption of energy which is a function of its wavelength (or frequency). The absorption profile, that is, the spectrum, is usually represented as a two-dimensional display in which the intensity of the absorption is measured along the ordinate and the frequency or wavelength of the incident radiation is recorded along the abscissa.

The classical description of the mechanism of interaction arises from a coupling of the electric vector of the incident radiation with the electric component of the dipole moment vector of a particular vibrating bond. Absorption of radiation occurs providing there is a change in the value of the dipole moment and provided that the incident radiation is of the same frequency as the equilibrium frequency of the bond.

In general, any group of complex vibrations, which most polyatomic molecules exhibit, can be resolved into several independent motions called "normal vibrations." If these vibrations involve changes in the local dipole moments, energy is absorbed from the incident radiation depending on its wavelength and the type of vibration. The sum total of these absorptions gives the spectrum of the molecule, and if the proviso is made that only vibrations are being considered, then the absorptions lie in the infrared region of the electromagnetic spectrum within the range of 0.7 to 1000 μm.

The infrared spectrum is often stated to be specific to a substance, as its normal vibrations represent a function of the geometric distribution of the vibrating nuclei and of the force field which tends to restore the molecule to its equilibrium configuration during distortion. This suggests that, in practice, identification of single substances is relatively easy because the molecular properties of shape and energy mentioned above are unique to each molecule. However, this view is an oversimplification and a cautious approach needs to be adopted as other factors have to be taken into account before an unambiguous identification can be made. The reasons for this are as follows: The vibrational spectrum of any molecule (excluding diatomics) will, in theory, contain (3 n-6) absorptions, where n is the number of atoms; thus 117 vibrations are possible in the molecule of the typical anionic detergent sodium lauryl sulfate ($C_{12}H_{25}OSO_3Na$). If harmonics are allowed, i.e., combinations and overtones, the number may be more. For small molecules and those having high symmetry the number of active vibrations is often much less than theory because the operation of selection rules allows for degeneracy of normal vibrations and also for inactive infrared modes. This does not usually apply in the case of larger molecules and the number of vibrations possible could easily reach over one hundred as shown above. A mathematical treatment of this situation, even if desirable, is impossible and consequently empirical methods must be used.

Fortunately it has been found that large molecules can be considered to be made up of independent submolecular units from the point of view of a vibrational analysis. This means that if the vibrational properties of simple molecules are known, then the transposition of these results to the identical groups in large molecules is valid. Rigorous mathematical analyses together with the compilation of a considerable amount of experimental data have provided a huge bank of information about the vibrational frequencies (and to a lesser extent about the intensities) of characteristic groups such as OH, CH_2, CN, C=O, and many others. Thus the carbonyl frequency in a long-chain ester falls in the same range as that of a simple ester, and so on. The assumption is made that the frequency values of common groups in large molecules are only slightly modified, if at all, by the rest of the molecule, which has given rise to the concept of the group frequency. This feature of the infrared method has, probably more than any other, led to its widespread use as an analytical tool. Tables of group frequencies in the infrared range are in all texts about the subject, and spectral profile matching via collections of infrared spectra is commonplace in most analytical laboratories.

B. Sampling and Sample Preparation

Samples range from the very pure individual chemical compound obtained, for example, as the eluate from a high-performance liquid chromatographic separation, through a mixture of homologous surfactants of the same class, to a mixture of various classes of anionics, to a commercially available detergent, and finally to a commercial formulation.

Useful infrared information can be obtained from all these samples but care must be taken when interpreting the data to always bear in mind the history of the sample being examined.

It is advisable to remove water from samples, as the presence of even traces of moisture can mask absorption in the 3400 to 3000 cm^{-1}, 1700 to

1500 cm^{-1}, and 600 cm^{-1} regions (due to the well-known high optical density of water). These frequency intervals contain useful information about the sample and clearly should be free from extraneous absorptions. In this context may be mentioned the need to remove inorganic substances, other types of surfactants (cationics, nonionics, amphoterics), and nonsurface-active organic materials. Separation methods are discussed in Chapter 1.

The physical state of the sample needs to be considered before selecting a particular method of preparation for study by the infrared technique. The most common method of sampling solids is by means of the alkali halide disc or by the formation of an oil mull, e.g., with the hydrocarbon Nujol or the fluorinated hydrocarbon Fluorolube. If the sample is a low-melting solid or a viscous fluid, a pressed-out film (POF) can be obtained between the appropriate transmitting plates. For the examination of a solution, various types of fixed and demountable cells are available which may be adjusted to give desired volumes, i.e., path length variability. In the case of anionic detergents, no unusual sampling methods have been employed and those listed above are extensively described in texts on the infrared method, e.g., by Miller and Stace [1] or Hummel [2]. If a satisfactory spectrum cannot be obtained by one sampling procedure, an alternative should be tried.

C. Presentation and Recording of the Spectrum

Spectra are usually presented as a two-dimensional display on a gridded chart in which the quantity measured along the ordinate is in units of intensity, most often transmittance (defined as 100 I/I_0) or absorbance or optical density (defined as $\log_{10} I_0/I$, where I and I_0 are the intensities of the transmitted and incident radiation, respectively) versus a wavelength or wavenumber (reciprocal wavelength) scale along the abscissa. Recently the trend has been to record wave number in preference to wavelength; absolute frequency (Hz) is hardly ever used as a practical measure of absorption characteristics. The wave-number presentation is advantageous because it is a direct measure of vibrational energy, band contours are symmetrical, and many regularities in the spectrum can be observed immediately. A slight drawback is the overemphasis of certain spectral regions, but this has been overcome by the use of a scale change at about 2000 cm^{-1}. There is some inconvenience when a standard spectrum, linear in wavelength, has to be compared with one that is linear in wave number, but the experienced operator usually overcomes this by familiarity with both methods of presentation.

D. Instrumentation

This wide-ranging topic is amply covered in several excellent texts specifically devoted to the chemical applications of infrared technology, e.g., Miller and Stace [1], Bellamy [3], Brugel [4], and Colthup et al. [5].

An important newer development in the rapid production of good quality infrared spectra has been the advent of the Fourier Transform (FT) instrument. This type of spectrometer accepts the sample in the same way as the conventional one but instead of scanning through the frequency range, usually 4000 to 400 cm^{-1}, over a period of time the sample is subjected to the whole frequency range simultaneously. The spectrum is produced in a fraction of the time and is generally of higher quality, i.e., better resolution. A major disadvantage is the cost, as the instrument requires a dedicated computer to unravel the Fourier Transform data and produce a standard spectrum. Despite the economic problem, the advantages in terms of improvement in signal-to-noise ratio for small quantities and other difficult samples (arising from the ability of the spectrometer to scan repeatedly and rapidly) has made the FT instrument a serious competitor to conventional IR spectrometers. The elementary theory and practice of the use of FT infrared spectrometers is outlined in the texts by Miller and Stace [1] and by Strobel [6].

E. Interpretation of the Spectra

This aspect of the subject is dealt with comprehensively by Miller and Stace [1], Bellamy [3], Colthup et al. [5], and others. A worthwhile addition to the standard reviews is the text by Hummel [2] which is devoted largely to the infrared analysis of surfactants.

Interpretation may be achieved by band assignment using the information found in the literature, or by comparison with known spectra. The most comprehensive reference system for infrared spectra is the Sadtler Standard Spectra Collection [7]. As of 1974, it contains 47,000 prism spectra and approximately 2000 are added each year. A collection of grating spectra is also available. In addition, there are collections of spectra of commercial compounds, one devoted to surface-active materials, which contains 5000 spectra [7].

F. Common Assignments of Functional Groups in
    Anionic Surfactant Molecules

The hydroxyl (OH) stretching frequency, arising from the hydroxyl group in carboxylic acids, covers the range 3400 to 2400 cm^{-1}. The band is usually complex, ill-defined and broad and these characteristics (attributed to hydrogen-bonding effects) are observed both in the solid and in solution. In the absence of H-bonding the OH stretching vibration absorbs at about 3500 cm^{-1}.

Superimposed on the low frequency side of the broad OH band are those absorptions arising from the C—H stretching of the methylene (CH_2) and methyl (CH_3) groups in the carbon chain, i.e., their symmetric and anti-symmetric vibrations, found in the range of 2950 to 2750 cm^{-1}. In the case of saturated soaps or fatty acids the position and shape of these absorptions are not noticeably different from other long-chain organic compounds and will not be considered further here.

For unsaturated materials some further information may be obtained. A C—H band due to the vibration associated with the C=C—H double bond can normally be identified at 3020 cm^{-1}, and its intensity relative to the methylene antisymmetric stretch at 2920 cm^{-1} gives an indication of the degree of unsaturation [8].

The carbonyl stretching vibration for monobasic saturated aliphatic acids, containing no electron-attracting substituents close to the carboxyl carbon atom and no H-bonding, is observed between 1725 and 1705 cm^{-1}. Flett [9] examined some 60 carboxylic acids, and all of the C=O absorptions fell in this range. Sinclair et al. [10] reported that saturated fatty acids dissolved in carbon disulfide have a strong carbonyl absorption at 1708 cm^{-1} associated with the dimer and a weaker band at approximately 1750 cm^{-1} probably due to the carbonyl absorption of the monomer. In the solid state, carboxylic acids exhibit a single band at 1708 cm^{-1} arising from the associated species.

The carbonyl absorption in unsaturated fatty acids could show a complex profile because of the weak absorption of the C=C double bond vibration in the 1650 to 1700 cm^{-1} range. Fortunately the C=C absorption is usually observed as a shoulder on the low-frequency side of the strong carbonyl band and it does not apparently alter the position of the band maximum which is still at 1708 cm^{-1}. In the case of cis-α,β-unsaturated acids, e.g., 2-octadecenoic, the double bond is conjugated with the carboxylic acid group, and as a result the carbonyl absorption drops to 1695 cm^{-1} with a shoulder at about 1650 cm^{-1} arising from the C=C vibration; both these bands have greatly enhanced intensities [8].

An aid to the analysis of long-chain carboxylic acids in the solid state (KBr discs) comes from the spectral profile in the range 1350 to 1180 cm^{-1}. Jones et al. [11] showed that these compounds gave rise to a progression of absorption bands of uniform spacing and intensity. Brown et al. [12] attributed this to vibrational interactions among the methylene groups, whereas Meiklejohn et al. [13] showed that the number of carbon atoms in the chain may be estimated from the number of bands, the regularity of the spacing, and their displacement to lower frequencies with increasing chain length. An empirical formula was found relating the number of bands to the number of carbon atoms, depending on whether the acid contained an even or odd number of carbon atoms in the chain (for C_{10} and above). Thus for compounds containing even numbers of carbon atoms in the chain:

$$\text{Number of bands} = \frac{\text{Number of C atoms}}{2}$$

For acids containing an odd number of carbon atoms in the chain there is a shift in the wavelength of the entire band group and the formula to be applied is:

$$\text{Number of bands} = \frac{\text{Number of C atoms} + 1}{2}$$

Branched-chain acids give similar overall spectral profiles but the absorption bands are not uniformly spaced and no simple formula leads to the determination of the chain length.

1. Soaps

Ionization of the carboxyl group of long-chain compounds gives rise to the formation of the carboxylate (COO^-) function. This results in the loss of the characteristic carbonyl absorption which is replaced by two absorption bands, a very strong one between 1610 and 1550 cm^{-1} (the antisymmetric C—O stretch) and another, usually much weaker, between 1410 and 1300 cm^{-1} (the symmetric C—O stretch). There may also be some other spectral changes because there is now no OH group present, but these may be less distinctive than changes in the 1750 to 1550 cm^{-1} region.

Childers and Struthers [14] made use of these changes in the analysis of the sodium salts of long-chain mono- and dibasic acids. The spectra, obtained from KBr discs or Nujol mulls over the range of 5000 to 600 cm^{-1} were specific to the salt, which was in contrast to the spectra of the acid form. In addition it was possible to obtain a quantitative estimation of the components in a mixture.

In nonaqueous solutions, fatty acid may be identified in the salt form by shifts in the carbonyl frequency as a function of metal type. However, it must be remembered that in these situations the metal is covalently bound to the acid and so the carbonyl group is altered only indirectly. Hence the absorption will vary with the metal present but within the range expected for the carbonyl frequency.

An infrared study of aluminum soaps was reported by Harple et al. [15], who showed that the mono- and dialuminum soaps existed as genuine chemical compounds and confirmed the absence of a trisoap. The monosoap could be distinguished readily from the disoap by differences observed in their infrared spectra; the former exhibited only a broad hydroxyl absorption whereas the latter showed a narrow, free OH absorption band at about 3600 cm^{-1}.

Chapman [16] showed that analysis of anhydrous sodium soaps depended on the frequency and number of bands in the 1350 to 1200 cm^{-1} region, the empirical rule mentioned previously for fatty acids holding in the case of their salts.

2. Anionic Detergents

Delsemme [17] showed how useful the infrared spectrometric technique could be in the analysis of anionic detergents by means of the functional-group frequency approach and later presented a paper on this topic at the first World Congress on Detergency [18]. In the meantime, Sadtler [19] reported on the identification of surfactants by infrared methods and suggested that under favorable conditions a specific compound may be identified.

The main frequencies of anionic surfactant functional groups are now well established and are listed in Table 1 together with some characteristic fatty acid and soap absorption bands.

TABLE 1

Frequencies of the Main Functional Groups in Anionic Surfactants, Soaps, and Fatty Acids

Functional group	Vibration	Frequency range, cm^{-1}	Intensity
OH	O—H stretch	3500–3100	strong to medium, broad
NH NH$_2$	N—H stretch	3300–3000	strong to medium, broad
CH	C—H stretch adjacent to double bond	3100–3000	
CH$_2$	C—H antisymmetric stretch	2936–2916	strong
	C—H symmetric stretch	2863–2843	weak
CH$_3$	C—H antisymmetric stretch	2972–2952	strong
	(aromatic)	(2930–2920)	
	C—H symmetric stretch	2882–2862	weak
	(aromatic)	(2870–2860)	
C=O	C=O stretch in esters	1750–1735	very strong

TABLE 1 (Cont.)

Functional group	Vibration	Frequency range, cm^{-1}	Intensity
C=O	C=O stretch in		
	fatty acids, monomer	1800–1740	very strong
	dimer	1720–1680	very strong
	soaps	1650–1550	weak
		1440–1370	weak
C=C	stretch	1680–1630	weak to medium
CH_2	deformations	1475–1450	medium
CH_3	antisymmetric deformation	1475–1450	medium
	symmetric deformation	1383–1377	medium
$-(CH_2CH_2O)-$, Ethylene oxide condensates		1250–1110	strong
$-C-SO_3$, sulfonate	S—O stretch		
	antisymmetric	1375–1335	very strong
	symmetric	1195–1165	medium
$C-O-SO_3$, sulfate	antisymmetric	1245–1213	very strong
	symmetric	1040–1010	medium
	S—O deformations	630	medium–strong
		620	medium–strong
		584	medium
CH_2	rocking $> (CH_2)_4$	720–726	weak
	$< (CH_2)_4$	726–785	weak

G. Applications

A group vibrational analysis of the barium salts of sulfates, sulfonates, ester sulfonates, and amide sulfonates was reported by Jenkins and Kellenbach [20] using the KBr pellet method. Characteristic absorption peaks enabled these different types of detergent to be easily identified.

La Lau and Dahmen [21] made organic solvent-soluble ion pair derivatives of commercial detergents which were subsequently examined by infrared spectroscopy. Bands were assigned to the internal vibrations of the sulfate group (OSO_3) between 1300 and 500 cm^{-1} which enabled a distinction to be made between alkyl sulfates on the one hand and alkyl- and arylsulfonates on the other.

The determination of alkylbenzenesulfonates in river waters has attracted increasing attention in recent years. The Analytical Subcommittee of the Association of American Soap and Glycerin Producers studied the problem and developed a method which finally led to infrared spectrometry [22]. Because of various interferences, colorimetric and titrimetric methods gave poor results and it was found necessary to extract the alkylbenzenesulfonate as an amine complex after a multistage separation procedure. The complex, the 1-methyl-n-heptylamine salt, was determined quantitatively by measuring its absorbance in carbon tetrachloride or carbon disulfide solution at 1042 and 1010 cm^{-1}. Good qualitative identification was obtained from the complex between sodium chloride plates and by observing absorptions in the ranges of 1250 to 1180 cm^{-1} (sulfonates), 1140 to 830 cm^{-1} (benzene ring modes), and 1397 to 1362 cm^{-1} (certain polypropylene chain vibrations).

A double extraction procedure was described by Fairing and Short [23] which was specific for the determination of alkylbenzenesulfonates (ABS) in sewage and surface water. Although the final assay, after the ABS was isolated, was accomplished by coupling with methylene blue and colorimetric determination, confirmation of the validity and accuracy of the data was obtained using infrared spectrometry. This was achieved by isolation of the ABS as its 1-methyl-n-heptylamine salt and examination of bands at 1042 and 1010 cm^{-1} in carbon disulfide solution.

When attention was focused on biologically "hard" and "soft" surfactants in sewage (the former are branched-chain and the latter linear-chain ABS compounds), infrared spectroscopy offered a good method for the discrimination and determination of these compounds, provided that extraction procedures were effective. Ogden et al. [24] made the n-heptylamine salt, dissolved it in carbon disulfide, and measured the absorbance of the band at 1010 cm^{-1} to estimate total ABS. The amount of "hard" ABS in the sample was found by measuring the band absorbance at 1370 cm^{-1} in carbon tetrachloride solution. The difference between total and "hard" ABS was a measure of the "soft" anionic surfactant present.

Further work in this area was done by Frazee and Crisler [25] who showed that the relative amounts of straight- and branched-chain ABS isomers in a mixture may be determined by comparing the infrared intensities of certain bands in their spectrum. Thus the ratio of the absorption intensity at 1367 cm^{-1} (the CH deformation in branched-chain CH_3 groups) to that at 1410 cm^{-1} (a band associated with the straight-chain sulfonate whose position depends on the nature of the alkyl group), was measured for the n-octylamine salts of ABS which were preferred to the 1-methyl-n-heptyl-amine derivatives since the methyl branch absorption in the analytical region was avoided. The ratio of straight-chain to branched-chain material was obtained from the observed ratio of intensities by reference to a calibration curve prepared from a known mixture of linear- and branched-chain alkylbenzenesulfonates.

Wright and Glass [26] claimed that to distinguish successfully between linear and branched-chain sulfur-containing detergents by infrared methods, desulfonation was necessary. Their argument was based on the fact that a weak band at 1190 cm^{-1} (attributed to the hydrocarbon chain) was obscured by the intense SO-stretching mode of the OSO_3 group. This band, together with one between 1390 and 1370 cm^{-1}, confirmed branching. A critical examination of their data does not seem to justify the effort of desulfonation since according to Ogden et al. [24] at least three bands can be used for the purposes of identification and determination of "soft" and "hard" detergents in the presence of each other (1412 cm^{-1} for "soft" and 1402 and 1422 cm^{-1} for "hard").

More recently Maehler et al. [27] combined and modified the earlier methods of Fairing and Short [23] and Frazee and Crisler [25] to differentiate between alkylbenzenesulfonates and the newer "soft" linear alkylsulfonates in underground waters. Their technique also permitted quantitative determination of the amount of each component to within ±10%.

A study on the analysis of alkylbenzenesulfonates present in sewage, using a modification of the technique of Ogden et al. [24], was reported by Hughes et al. [28]. The method consisted of the empirical determination of the degree of hardness of the isolated sulfonates by comparison of the intensity of the absorption of the n-heptylammonium salts at 1368 cm^{-1} ("hard") with that at 1378 cm^{-1} ("soft" and "hard"). The results obtained for "hard" detergents were quoted in terms of percentage tetrapropylenebenzenesulfonate, using calibration data from synthetic mixtures. The lower limit of detection of branched-chain material was estimated at 3%.

Matsumoto et al. [29] determined "hard" and "soft" detergents in household products using the infrared absorption of solutions in carbon disulfide at 673 and 659 cm^{-1}, respectively, as the final step in this assay. Results were obtained from samples converted to the n-heptylamine salt after either cation exchange or petroleum ether extraction following acid hydrolysis. Comparison of the results showed that the standard deviation for the solvent extraction method (1.4%) was less than that for the cation exchange (2.2%).

A simple infrared technique for detergent level determination suitable for a quality control laboratory was devised by Brunelle and Crecelius [30]. It is based on the method of determination of hydroxyl groups used by Burns and Muraca [31] in their studies of polypropylene glycols. The method involves careful calibration of the hydroxyl absorption of a chosen alcohol as a function of concentration, to which the unknown alcohol (i.e., hydrolyzed detergent) is referred. The procedure is fast but has some inherent disadvantages; for example, no amine or water must be present; in addition, a hydrolysis step is essential for detergent analysis and clearly a preseparation stage must be included.

Abe and his co-workers [32-34] studied the analysis of certain components in household detergents, including infrared methods. They established the limit of detection of such materials as p-toluenesulfonate at about 5% from an absorption band at 805 cm^{-1}, and of sodium stearate about 2% from the band at 1650 cm^{-1}. Many other materials were also examined.

In the case of liquid detergents, an infrared study was made to examine the effect of changes in the counter ion. The samples were first dried under vacuum and then examined as halide discs. A typical result showed that the band at 1045 cm^{-1} in n-alkylbenzenesulfonates and isoalkylbenzenesulfonates shifted to lower frequencies on changing the counter-ion through the series ammonium, mono-, di-, and triethanolamine [33]. By carrying out a series of experiments with different detergents and observing individual shifts with counter-ion it was possible to identify qualitatively the detergent in household products. However, if urea was present, it interfered with the spectrum and made identification difficult.

Mixtures of detergent powders (alkyl sulfate and alkylbenzenesulfonate) were analyzed with IR spectroscopy, using the KBr technique, after extraction of the material with ethanol [34]. Qualitatively alkyl sulfate was distinguished from benzenesulfonate by strong absorptions at 1085 and 1000 cm^{-1}; the limit of detection was estimated at 10%. Quantitatively the amount of alkyl sulfate was found by measuring the fractional change in extinction coefficient based on the formula

$$\frac{\epsilon_{1085 \ cm^{-1}}}{\epsilon_{1085 \ cm^{-1}} + \epsilon_{1045 \ cm^{-1}}}$$

where the 1045 cm^{-1} band comes from the alkylbenzenesulfonate. Exact determination of the material in the mixture depended on the procedures for selecting a baseline from the spectrum which, in turn, depended on whether the alkyl sulfate content was more or less than 40%. These results were compared with data obtained by the hydrolysis method of House and Darragh [35]. The standard deviation in the quantitative analysis of alkyl sulfate in mixtures was 1.81 and 1.76% for the IR and the hydrolysis methods, respectively.

Saiki [36] analyzed for minor components, such as ABS, in SBR latex, a compounding agent for Portland cement. The latex was extracted with methanol and any surfactant in the extract identified by infrared techniques.

Hashimoto et al. [37] identified three classes of anionic detergent by means of the characteristic absorption bands of their corresponding sulfonyl chloride derivatives. These were at 655 cm^{-1} for alkyl sulfates, 524 cm^{-1} for α-olefinsulfonates, and 640 cm^{-1} for linear alkylbenzenesulfonates.

Nettles [38] reviewed the usefulness of infrared spectroscopy in the analysis of surfactants frequently encountered by textile chemists. He discusses in detail the spectra of individual members of several classes of surfactant, including fatty acid soaps, ethylene oxide derivatives, sulfuric acid esters, and others.

The application of attenuated total reflectance (ATR) infrared spectroscopy to the problem of qualitative identification of components in toiletry products was reported by Puttnam et al. [39]. This applied infrared technique is now well established and frequently used to study samples which would be difficult or impossible to examine by conventional transmission IR. Essentially the reflectance method is independent of sample thickness as the interactive radiation only penetrates a short distance into the sample; even at long wavelengths it is rarely greater than 5 μm. The main advantage of ATR is that the sample may often be examined directly without any preparation. By this means it is possible to rapidly identify band absorptions related to anionic, soap, and fatty acid material, particularly when they are present as surface layers on solid substrates.

II. IDENTIFICATION BY ULTRAVIOLET
SPECTROSCOPY

Molecules containing aromatic nuclei or those which are unsaturated in any way exhibit characteristic absorptions in the ultraviolet region of the spectrum (i.e., 200–400 nm). Thus anionic surfactants containing groups of this type may be identified from their UV absorption characteristics.

A. Sampling and Sample Preparation

Most anionic surfactants are soluble in water or lower alcohols and any of these solvents are suitable vehicles for the determination of UV spectra. The presence of other substances may interfere and it is advisable to eliminate possible interferences before UV examination. The sample solution can be diluted or concentrated as required to suit the parameters of the instrument.

B. Recording the Spectrum

Any medium-to-high resolution single- or double-beam recording UV spectrometer is suitable. Unless a specific absorption region is of interest, it is advisable to record the spectrum from 210 to 350 nm. It is not usually necessary to record spectra below 210 nm.

C. Interpretation of the Spectrum

The sample spectrum can be compared with reference spectra, preferably run on the same instrument using similar conditions and solvents.

A limited number of anionic surfactants are included in the Sadtler Standard Spectra series [40] of UV spectra, but there is no equivalent to the commercial infrared spectra series of surfactants to cover the UV region.

D. Applications

Reid et al. [41] illustrated the utility of UV spectrometry for the qualitative identification of anionic surfactants, and recommended preliminary separation of the anionic from other compounds. They showed that unsaturated soaps (e.g., oleates) could be distinguished from saturated soaps (e.g., stearates), and that alkylarylsulfonates could be distinguished from alkyl sulfates. It was also possible to differentiate between alkylbenzene-, alkylnaphthalene-, tetrahydronaphthalene- and arylbenzenesulfonates. Aniline derivatives (e.g., dibenzylsulfanilates) could also be identified.

Kelly et al. [42] used the absorption at 224 nm for the rapid determination of alkylarylsulfonates in the presence of other detergents for production control. The sulfonates were continuously monitored using a flow cell with a path length of 0.03 cm.

Cullum [43] indicated that measurement of the absorption band maximum at 224 nm can be used as a very sensitive method for the determination of alkylbenzenesulfonates at concentrations of only a few ppm.

Izawa [44] found a linear relationship between extinction and percentage concentration for sodium alkylbenzenesulfonate at the 261 nm maximum and for the formaldehyde condensate of naphthalenesulfonate at 289-290 nm.

Arpino and de Rosa [45] determined sodium alkylarenesulfonates directly in sulfonation pastes or mixtures. They cite examples, calculate extinction coefficients, and measure the differences between various commercial products as well as show how interferences due to the unsulfated portion, inorganic salts, and organic additives can be evaluated.

Miranova and Malysheva [46] obtained spectra for a range of alkylbenzenesulfonates and found that the UV spectra could be used to obtain both qualitative and quantitative results. They determined the type of aromatic nucleus in detergents and the number of substituents in the benzene ring of alkylbenzenesulfonates.

Bartha et al. [47] showed that the UV absorption spectra of dodecylbenzenesulfonate at 225 nm remains unchanged in the presence of alkyl sulfates, except for an upward shift from the baseline. By using the extinction maxima of the aromatic ring they could determine the amount of alkylbenzenesulfonate in a mixture. The difference in extinction maxima at 225 and 270 nm was related to the concentration. Beers law was obeyed but the curve did not pass through the origin. The relationship, calculated from the average of the calibration curves, was

$$c = (0.0518 \pm 0.00184) + (0.00408 \pm 0.0045)C_d$$

where c = the difference in extinction maxima at 225 and 270 nm and C_d = the percentage of sodium dodecylbenzenesulfonate.

Ziolkowsky [48] also reported that the presence of benzene or naphthalene constituents could be detected by examination of the UV spectra. Straight-chain alkylbenzenesulfonates have a characteristic absorption band at 223 nm which is assigned to the benzene ring. The intensity of the absorption at this wavelength is always reduced after biodegradation, confirming that the aromatic ring is degraded in this process.

Tokiwa and Moriyama [49] mixed solutions of surfactants with and without benzene rings and measured the intensity of the UV absorption bands. Addition of sodium dodecyl sulfate to sodium benzenesulfonates with C_{10} and C_{12} side chains produced virtually no change in the position of the absorption band of the aromatic surfactants at 261 nm but there was a slight increase in the molar absorptivity. Addition of a polyoxyethylene (9.0 EO) dodecyl ether to the same sodium benzenesulfonates again produced no change in the position of the absorption band but the molar absorptivity was markedly decreased. This was thought to be caused by vibrational distortion of the benzene ring in the molecule.

ACKNOWLEDGMENT

The authors are indebted to Dr. C. B. Baddiel for his assistance and contribution to the infrared section of this chapter.

REFERENCES

1. R. G. J. Miller and B. C. Stace (eds.), Laboratory Methods in Infrared Spectroscopy 2nd Edition, Heyden and Son Ltd., London, 1972.
2. D. Hummel, Identification and Analysis of Surface-active Agents by Infrared and Chemical Methods, Interscience Publishers, New York, 1962.
3. L. J. Bellamy, The Infrared Spectra of Complex Molecules 2nd Edition, Methuen and Co. Ltd., London, 1958.
4. W. Brugel, An Introduction to Infrared Spectroscopy, Methuen and Co. Ltd., London, 1962.
5. N. B. Colthup, L. H. Daly, and S. E. Wiberley, Introduction to Infrared and Raman Spectroscopy, Academic Press, New York, 1964.

6. H. A. Strobel, Chemical Instrumentation: A Systematic Approach 2nd Edition, Addison-Wesley Publishing Co., Reading, Mass., 1963.

7. Sadtler Spectra Collections—Standard, Grating, Commercial, Heyden and Son Ltd., London, updated annually.

8. R. G. Sinclair, A. F. McKay, G. S. Myers, and R. N. Jones, The infrared absorption spectra of unsaturated fatty acids and esters. J. Am. Chem. Soc. 74, 2578-2585 (1952).

9. M. St. C. Flett, The characteristic infrared frequencies of the carboxylic acid group. J. Chem. Soc., Part 2, 962-967 (1951).

10. R. G. Sinclair, A. F. McKay, and R. N. Jones, The infrared absorption spectra of saturated fatty acids and esters. J. Am. Chem. Soc. 74, 2570-2575 (1952).

11. R. N. Jones, A. F. McKay, and R. S. Sinclair, Band progressions in the infrared spectra of fatty acids and related compounds. J. Am. Chem. Soc. 74, 2575-2578 (1952).

12. J. K. Brown, N. Sheppard, and D. M. Simpson, The interpretation of the infrared and raman spectra of the normal paraffins. Phil. Trans. Roy. Soc., London A247, 35-38 (1954).

13. R. A. Meiklejohn, R. J. Meyer, S. M. Aronovic, H. A. Schuette, and V. W. Meloch, Characterization of long-chain fatty acids by infrared spectroscopy. Anal. Chem. 29, 329-334 (1957).

14. E. Childers and G. W. Struthers, Infrared evaluation of sodium salts of organic acids. Anal. Chem. 27, 737-741 (1955).

15. W. W. Harple, S. E. Wiberley, and W. H. Bauer, Infrared spectra of aluminum soaps. Anal. Chem. 24, 635-638 (1952).

16. D. Chapman, An infrared spectroscopic examination of some anhydrous sodium soaps. J. Chem. Soc., 784-789 (1958).

17. A. H. Delsemme, Infrared analysis for functional groups in surfactants. Mededel. Vlaam. Chem. Ver. 13, 152-158 (1951).

18. A. H. Delsemme, Analysis and identification of surfactants by infrared spectrometry. Proc. Intern. Congr. Surface Activity, 1st, Paris, Vol. 1, 1954, pp. 192-196; C.A. 51, 17203b.

19. P. Sadtler, Analysis of synthetic detergents by infrared absorption technique. ASTM Bull., No. 190, 51-53 (1953).

20. J. W. Jenkins and K. O. Kellenbach, Identification of anionic surface-active agents by infrared absorption of the barium salts. Anal. Chem. 31, 1056-1059 (1959).

21. C. La Lau and E. A. M. F. Dahmen, Infrared study of some organic salts in solution—cyclohexylammonium salts of detergents. Spectrochim. Acta, 594-600 (1957 suppl.).

22. E. M. Sallee, J. D. Fairing, R. W. Hess, R. House, P. W. Maxwell, F. W. Melpolder, F. M. Middleton, J. Ross, W. C. Woelfel, and P. J. Weaver, Determination of trace amounts of alkylbenzenesulfonates in water. Anal. Chem. 28, 1822-1826 (1956).

23. J. D. Fairing and F. R. Short, Spectrophotometric determination of alkylbenzenesulfonate detergents in surface water and sewage. Anal. Chem. 28, 1827-1834 (1956).
24. C. P. Ogden, H. L. Webster, and J. Halliday, Determination of biologically soft and hard alkylbenzenesulphonates in detergents and sewage. Analyst 86, 22-29 (1961).
25. C. D. Frazee and R. O. Crisler, Infrared determination of alkyl branching in detergent ABS. J. Am. Oil Chemists' Soc. 41, 334-335 (1964).
26. E. R. Wright and A. L. Glass, Infrared analysis of detergents. Soap Chem. Specialities 41, 59-62, 83-84 (1965).
27. C. Z. Maehler, J. M. Cripps, and A. E. Greenberg, Differentiation of LAS and ABS in water. J. Water Pollution Control Federation 39, R92-R98 (1967); C.A. 68, 71979r.
28. W. Hughes, S. Frost, and V. W. Reid, Analysis of alkylbenzenesulfonates present in sewage. Proc. Intern. Congr. Surface Activity, 5th, Barcelona, Vol. 1, Ediciones Unidas S.A. Barcelona, 1968, pp. 317-325.
29. I. Matsumoto, H. Hasegawa, S. Togano, and K. Ishiwata, Infrared absorption spectrometry of sodium alkylbenzenesulfonates. Kogyo Kagaku Zasshi 72, 549-552 (1969); C.A. 71, 23147h.
30. T. E. Brunelle and S. B. Crecelius, Surfactant analysis—Infrared analysis applied to hydroxyl number determination. Soap Chem. Specialities 39, 63-65, 121 (1963).
31. E. A. Burns and R. F. Muraca, Determination of hydroxyl concentration in propylene glycols by infrared spectroscopy. Anal. Chem. 31, 397-399 (1959).
32. K. Abe, S. Tanimori, and S. Hashimoto, Composition analysis of detergents. V. Application of IR to the analysis of household detergents. Kogyo Kagaku Zasshi 69, 632-637 (1966); C.A. 65, 17211g.
33. K. Abe, H. Onzuka, and S. Hashimoto, Analysis of compounds in detergents. IX. Application of IR to analysis of surfactants in liquid detergents. Kogyo Kagaku Zasshi 69, 2019-2021 (1966); C.A. 66, 77265.
34. K. Abe, S. Tanimori, and S. Hashimoto, Analysis of compounds in detergents. X. Analysis of sodium alkyl sulfates in detergent powders. Bunseki Kagaku 15, 1364-1368 (1966); C.A. 66, 106173m.
35. R. House and J. L. Darragh, Analysis of synthetic anionic detergent compositions. Anal. Chem. 26, 1492-1497 (1954).
36. Y. Saiki, Infrared absorption spectral analysis of minor components in SBR latex as the compounding agent of Portland cement mortar. Bunseki Kagaku 15, 864-866 (1966); C.A. 66, 105728j.
37. S. Hashimoto, H. Tokuwaka, and T. Nagai, Determination of olefin sulfonates, linear alkylbenzenesulfonate and alkyl sulfates in detergents by infrared spectroscopy. Bunseki Kagaku 22, 559-563 (1973); C.A. 79, 80638h.

38. J. E. Nettles, Infrared spectroscopy for identifying surfaetants. Text. Chem. Color 1, 430-441 (1969).

39. N. A. Puttnam, S. Lee, and B. H. Baxter, Application of attenuated total reflectance IR spectroscopy to toilet articles and household products. 1. Qualitative analysis. J. Soc. Cosmet. Chem. 16, 607-615 (1965).

40. Sadtler Standard Spectra—Collection of Ultraviolet Spectra, Heyden and Son Ltd., London, updated annually.

41. V. W. Reid, T. Alston, and B. W. Young, The qualitative analysis of surface-active agents. Analyst 80, 682-689 (1955).

42. R. M. Kelly, E. W. Blank, W. E. Thompson, and R. Fine, Determination of alkylarylsulfonates by UV absorption. ASTM Bull., No. 237, 70-73 (1959).

43. D. C. Cullum, Wanted: A better method of analysing detergents. Mfg. Chemist 33, 352-356, 360 (1962).

44. Y. Izawa, Analysis of surface-active agents. XIX. Ultraviolet absorption spectra of aromatic surfactants. Yukagaku 11, 627-630 (1962); C.A. 58, 10409d.

45. A. Arpino and V. de Rosa, Analytical chemistry of surfactants. II. Ultraviolet spectrophotometric determination of sodium alkylarenesulfonate. Riv. Ital. Sostanze Grasse 39, 386-393 (1962); C.A. 60, 9487c.

46. A. N. Miranova and L. A. Malysheva, Spectral analysis of ABS. II. Analysis from UV absorption spectra. Tr. Vses. Nauchn.-Issled. Inst. Zhirov 23, 326-335 (1963); C.A. 63, 804g.

47. B. Bartha, I. Zambo, and G. Palyi, Determination of surface-active agents in the presence of each other. I. Determination of sodium dodecylbenzenesulfonate in mixtures with sodium fatty alcohol sulfates. Kolor Ert. 8, 384-385 (1966); C.A. 66, 106168p.

48. B. Ziolkowsky, Spectroscopy in the visible and the ultraviolet. II. Various practical application possibilities of ultraviolet spectroscopy. Seifen, Öle, Fette, Wachse 93, 99-102 (1967).

49. F. Tokiwa and N. Moriyama, Ultraviolet spectral study of mixed surfactant solutions. Nippon Kagaku Zasshi 91, 903-906 (1970); C.A. 74, 81375d.

Chapter 3

GAS CHROMATOGRAPHY FOR THE ANALYSIS OF ANIONIC SURFACTANTS

Toyozo Uno and Terumichi Nakagawa

Faculty of Pharmaceutical Sciences
Kyoto University
Kyoto, Japan

I. INTRODUCTION

Gas chromatography (GC) is a technique based upon the principle that each component of a mixture undergoes its own characteristic partition between the gas phase and a stationary (liquid or solid) phase packed in a column. The degree of separation of the sample components is determined mainly by the partition coefficients of the substances (solutes). These in turn are functions of the solute interaction with the stationary phase and of the vapor pressure of the solute at the column temperature. There are some other factors which affect the separation of the solutes and the shape of the subsequent chromatogram, such as solute diffusion, carrier-gas velocity, mass-transfer effect, etc., but these are not discussed here. It will be readily appreciated that the sample components require full vaporization prior to the partition so that the carrier-gas flow can drive them into the column; this is the first step of gas chromatography. Then one may suppose that the higher the volatility of the sample, the easier the gas chromatography and in general this is true. In this connection, it is logical that gas chromatography was first applied to the analysis of volatile materials such as petroleum products and only later to that of organic substances of little or no volatility such as drugs and polymers, and of some inorganic compounds. In the case of substances of little or no volatility (which includes surfactants), the samples require pretreatment by chemical reaction or thermal decomposition in order to permit GC analysis.

Surfactants, containing both hydrophilic and hydrophobic (lipophilic) groups in one molecule, have been traditionally classified in terms of the ionic properties of the hydrophilic groups into anionics, cationics, ampholytics, and nonionics. Such a classification suggests that the main interest so far has centered on the ionic properties of the surfactants. Accordingly, most analytical work has been aimed at quantitative and qualitative analysis of the hydrophilic groups and techniques for the analysis of hydrophobic groups have received less attention until recently. This may partly be due to the fact that there have been few powerful methods for the separation and identification of complex mixtures of the hydrophobic groups of surfactants, although paper, thin-layer, and column chromatographies and electrophoresis have been capable of separating them with relatively poor results.

It is well known that physicochemical and biological properties of surfactants such as affinity to vital tissues, biodegradability, critical micelle concentration (CMC), detergency, etc., are related to the chemical structure of the hydrophobic group. Above all, in respect to the current problem of water pollution, it became important to clarify the relationship between biodegradability of surfactants and chemical structure of hydrophobic groups. This seems to be one of the reasons why analysis of hydrophobic groups has made remarkable progress during the last 15 years.

Gas chromatography has played an important role in this progress; it is an important application of GC to separate complex mixtures, especially

of alkyl homologues and their isomers. However, there is a difficulty in
the practical application. As mentioned above, samples need to be vaporized
before the separation in the column, and surfactants have to be converted
to characteristic nonsurfactive hydrophobic oils which can be subjected to
GC analysis. Some nonionic surfactants can be separated without conversion to the hydrophobic oils, but these are not discussed in this chapter,
which describes chemical pretreatments and thermal decompositions followed by gas-chromatographic separation and determination of hydrophobic
parts of anionic surfactants. The separation of anionic surfactants from
other chemical substances and the separation of different types of hydrophilic groups from each other are not also included here since they are outside the realm of GC analysis and are discussed elsewhere.

Most anionic surfactants have a sulfonate, sulfate, or carboxylate
group as the hydrophilic part, and alkyl, alkylaryl, or alkenyl groups as
the hydrophobic part. The chemical bonds combining these two parts are
C—S, C—O, or C—C, respectively. The pretreatments, therefore, include
the severance of such bonds by making effective use of their characteristic
reactivities. In the following sections chemical pretreatments, including
thermal decomposition, gas-chromatographic determination of alkyl carbon
number and isomer distributions of anionic surfactants, are discussed in
detail after subdivision according to the conventional classification of their
hydrophilic groups.

II. ALKYLBENZENESULFONATES

Alkylbenzenesulfonate (ABS), usually the sodium salt, is one of the most
important anionic surfactants. There are two types of hydrophobic groups
of ABS leading to the so-called biologically "soft" and "hard" surfactants.
As the former is usually made by the cracking of higher paraffins and contains a high proportion of straight alkyl chain in its structure, it is relatively easily decomposed by bacteria dwelling in water. On the other hand,
the hydrophobic group of ABS, originating from a synthetic source (e.g.,
propylene polymers) contains highly branched alkyl chains that are resistant
to bacterial degradation and cause serious water pollution. Such a difference in the chemical structure of hydrophobic groups can be easily distinguished from their characteristic gas chromatogram, which can be obtained
by the methods discussed in the following sections.

A. Acid Decomposition

The chemical pretreatment for recovering the hydrophobic oil from ABS
utilizes the reactivity of the sulfonates with such inorganic acids as phosphoric, sulfuric, and hydrochloric acid. It is well known that aromatic
sulfonates are desulfonated at elevated temperatures to yield characteristic

oils [1, 2], the reaction generally being accomplished by treatment with
60-70% sulfuric acid and superheated steam at 140-190°C, or with concentrated hydrochloric acid in a sealed container at 150-200°C. However, use
of phosphoric acid [3-5] is preferable since it enables the reaction to pro-
ceed at atmospheric pressure with less carbonization than caused by sulfuric acid. Ease of desulfonation varies with the type of sulfonate, polyal-
kylbenzenesulfonate being much more reactive than monoalkylbenzenesulfo-
nate. Knight and House [6] applied this reaction to the pretreatment of
various types of surfactant and hence succeeded in the GC analysis of hydrophobic groups of ABS. Their procedure for recovering the hydrophobic oil
is as follows:

> Heat about 5 g of sample in 250 ml 93% phosphoric acid for 90 min
> at 215°C in a specially designed glass vessel (Fig. 1) and transfer
> the mixture of desulfonated products and water collected in the

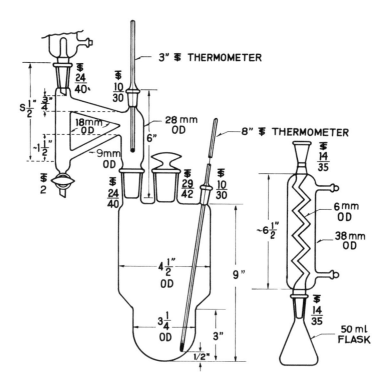

FIG. 1. Apparatus for decomposition of ABS with phosphoric acid.
Reprinted from Ref. 6 by courtesy of the American Oil Chemists' Society.

trap to a 125 ml separatory funnel using 5 ml acetone and 5 ml water. Add acetone until the aqueous layer contains 70% v/v acetone,
and then 20% aqueous potassium hydroxide until the mixture is neutral to Gramercy indicator (i.e., blue). Extract the hydrophobic
oil from the solution with three portions (10 + 10 + 5 ml) of isopentane, wash the combined extracts twice (10 + 10 ml) with water and
filter. For a complete recovery, rinse the funnel and filter paper
with a small quantity of isopentane. The solution of the neutral
hydrophobic oil is then submitted for analysis.

From GC, mass spectrometry (MS), and infrared (IR) measurements,
it was found that the main component of the hydrophobic oil recovered from
ABS is the precursor of ABS, namely alkylbenzene (AB). The yield of recovered oil increased with increasing reaction temperature from 1-2% theoretical at 185°C (90 min reaction time) up to about 80% at 215°C (90 min)
(see Table 1). The carbonization rate also increased with the temperature

TABLE 1

Effect of Certain Variables on the 90-Min Yield of Oil from
Commercial Sodium Dodecylbenzenesulfonate (348 Avg. MW)[a]

Sulfonate charged, g	Temp., °C	Additive, g	Recovered oil, yield, % theoretical
1	185	—	1-2
1	190	—	25
1	200	—	49
2	215	—	78
11	215	—	78
1	224	—	49
2[b]	215	—	68-71
2	215	Sodium sulfate, 2	79
2	215	Sodium sulfate, 5	79
2	215	Sodium sulfate, 1	82

[a]Reprinted from Ref. 6 by courtesy of the American Oil Chemists' Society.
[b]Sulfonate charged as the complex formed from 20 g of Dowex 1-X2 ion exchange resin plus 2 g of the sodium sulfonate.

so that the yield of alkylbenzene was less at 225 than at 215°C. This was true for commercial ABS as well as synthetic monoalkylbenzenesulfonates of known structure. The rate of desulfonation increased with the amount of the sulfonate in the reaction mixture during the first 60-90 min and also with the decreasing molecular weight of the alkyl group. From these preliminary examinations, 215°C, 90 min, was found to give the best yield of hydrophobic oil. As Knight and House mentioned, little change was found in either molecular weight or structure of monoalkylbenzene, GC and MS analyses indicating a small amount (less than 1%) of degradation to lower boiling materials, and IR spectra indicating the formation of small amounts of olefinic or oxidation products. These by-products account for the slight difference between chromatographic patterns of desulfonated ABS and raw material of the same ABS, and suggest that highly branched alkyl chains, if present, are preferably dealkylated during the desulfonation reaction.

Nishi [7], attributing such undesirable dealkylation to the oxidizing power of sulfuric acid formed from the sulfonate group of ABS during the reaction, has modified the method by adding metallic tin or stannous chloride to the reaction mixture in order to reduce the oxidizability: this trial was successful in the case of short-chain ABS, but was regrettably not extended to detergent-range sulfonates. Some other modifications have also been presented. For example, a microdesulfonation technique [8] (less than 10 mg of sample) with refluxing the sulfonate in a specially designed apparatus with phosphoric acid for 60-90 min; heating with concentrated hydrochloric [9] instead of phosphoric acid in a sealed tube at 200°C for 3 hr; and desulfonation in a pressure tube [10, 11] with phosphoric acid at 250°C for 15 min.

The acid-decomposition methods, on the whole, are not ideal for the pretreatment of ABS for GC analysis but have contributed by initiating the development of on-line acid-pyrolysis gas chromatography of ABS (Sec. II. D).

As mentioned above, alkyl carbon number and isomer distribution for the hydrophobic group of ABS vary widely with the origin of the raw material. Hard ABS has a far more complicated distribution than soft ABS; C_{10}-C_{15} propylene polymer-derived ABS, o- and p-isomers inclusive, could theoretically contain up to about 80,000 possible species [12], and linear ABS only 74. Figure 2 shows typical chromatograms for the precursors of both types of ABS [13]. The peaks due to tetrapropylene-derived phenyldodecane are incompletely resolved in a relatively narrow range, suggesting that this AB was obtained by the narrow-cut distillation of propylene polymers. The straight-chain AB yields a countable number of well-resolved peaks which are more widely distributed than those obtained from hard ABS. In this case, the peaks emerge as usual in the order of alkyl carbon number (from low to high) for homologous series and substituent (phenyl group) position (from inner to outer) for positional isomers. The high resolving power of gas chromatography has, therefore, significant application in fields where the hydrophobic group distribution plays an important role.

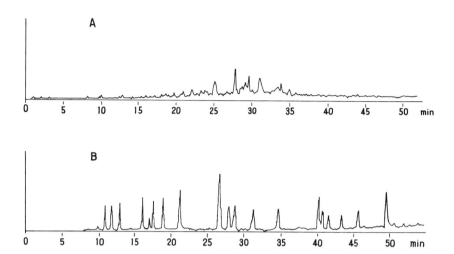

FIG. 2. A. Chromatogram of tetrapropylene-derived phenyldodecane; 50 m × 0.25 mm stainless-steel capillary column coated with DC-550 (polar methylphenyl silicone oil), 120-170°C programmed at 1.5°C/min. B. Chromatogram of a straight-chain alkylbenzene; 50 m × 0.25 mm stainless-steel capillary column coated with Apiezone L, 150-220°C programmed at 2°C/min after first 30 min. Reprinted from Ref. 13 by courtesy of the American Chemical Society.

Swisher [14] employed the phosphoric acid decomposition method during his investigation of the mechanism of the biodegradation (β-oxidation) of ABS and obtained significant information correlating biodegradability with the chemical structure of the hydrophobe. Jungermann et al. [15] also used this method with little modification to investigate the relationship between the gas-chromatographic pattern of desulfonated products of ABS and its detergency.

B. Alkali Fusion

The alkali-fusion reaction, characteristic of aromatic sulfonates, may also be used for the chemical pretreatment of ABS. It is well known that aromatic sulfonates are converted to the corresponding phenols by this reaction [16], e.g., for benzenesulfonate:

$$C_6H_5SO_3Na + 2\ NaOH \longrightarrow C_6H_5ONa + Na_2SO_3 + H_2O$$

Nishi [17] utilized this reaction for the pretreatment of ABS and obtained alkylphenols as the hydrophobic oil. His procedure is as follows:

Mix 10-100 mg of sulfonate with 500 mg of KOH in a glass tube
(8 mm × 10 cm) and heat for 5 min at 350-360°C in a silicone-oil
bath while stirring with a thin glass rod. Cool, dissolve the contents in 2 ml of water and neutralize with conc. HCl using Congo
red as an indicator. Extract the hydrophobic oil with 0.5 ml of
benzene.

The gas chromatogram of the oil thus obtained showed numerous peaks due
to alkylphenols under the column conditions employed. Figure 3 shows the
alkali-fusion gas chromatograms of some soft and hard ABS in which the
chromatographic pattern of alkylphenols (solid line) dependent upon molecular weight distribution, degree of branching of alkyl chain and position of
substitution on the benzene ring, shows high similarity to that of the original
alkylbenzene (dotted line). Such correlation can facilitate the discrimination of hydrophobic groups of ABS, although all the peaks are not identified
with authentic samples. This method has another advantage because it is
possible to separate the positional isomers of ABS, i.e., o- and p-alkyl
substituted benzenesulfonates if improved GC conditions are applied (e.g.,
use of capillary columns).

C. Conversion to Sulfonyl Chlorides

Kirkland [18] reported conversion of nonvolatile sulfonic acids and their
salts to sulfonyl chlorides, basing his method upon the reaction of aromatic,
aliphatic, and alkylarylsulfonic acids and their salts with thionyl chloride
or with phosgene [19] in the presence of dimethyl formamide.

$$RSO_3H + SOCl_2 \longrightarrow RSO_2Cl + SO_2 + HCl$$

Yields of 96-100% were obtained for all the sulfonates examined, but unfortunately, sulfonyl chlorides generally suffer thermal decomposition to some
extent [20] at the temperature of the injection port and column-oven of the
gas chromatograph. This occurred also in the case of sulfonic acid methyl
esters and a reduced-pressure gas chromatographic technique (inlet pressure of about one-half to one-third of atmospheric) was employed to vaporize
the sample before significant degradation could occur [18]. Considering
these facts, the application of this reaction to ABS seems to require special
precautions because the sulfonyl chlorides derived from detergent-range
ABS have low volatilities and therefore need higher temperatures for vaporization. Such thermal instability, however, facilitates pyrolysis gas chromatography of alkylsulfonates (Sec. IV).

Parsons [21], attempting to increase the volatility, converted sulfonyl
chlorides obtained by using PCl_5 (instead of $SOCl_2$) to sulfonyl fluorides,
which are in general more volatile and stable than the corresponding

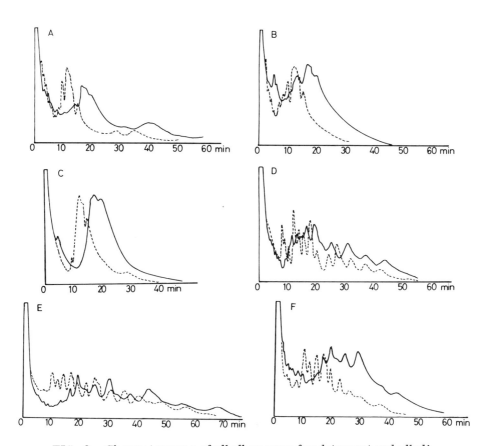

FIG. 3. Chromatograms of alkylbenzenes for detergent and alkali fusion products of their sulfonate. Key: ——— Alkali fusion product (2% SE-30 on Chromosorb W, 3 m × 4 mm i.d., 150°C, 60 ml/min, flame ionization detector; ----- Original alkylbenzene (1% in benzene) 2% SE-30 on Chromosorb W, 3 m × 4 mm i.d., 130°C, 60 ml/min, flame ionization detector. A. Alkane 60 N; B. Alkane 56N; C. Mitsubishi 253; D. Conoco N600; E. Conoco N500; F. Dobane J.N. Reprinted from Ref. 17 by courtesy of the Japanese Society of Analytical Chemistry.

chlorides and can be gas chromatographed at relatively lower temperatures, but there has been no report to date concerning application of this technique to detergent-range ABS. Another attempt to lower the vaporizing temperature and to increase the volatility was made by reducing the sulfonyl chloride to thiol with LiAlH$_4$, followed by conversion to the trimethylsilyl (TMS) derivative by use of BSA (bistrimethylsilyl acetamide) [22]. This is a rare example of direct sililation of a sulfur atom, but details are not yet published. Further investigation of this reaction should lead to its application in the pretreatment of ABS for GC analysis.

D. On-line Pyrolysis

The application of direct on-line pyrolysis gas chromatographic technique was first reported by Liddicoet et al. [23], who decomposed ABS at 650°C

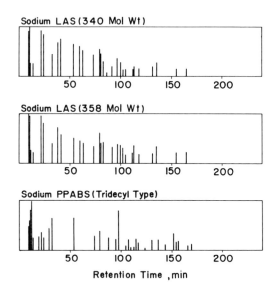

FIG. 4. Pyrolysis gas chromatograms of alkylbenzenesulfonates. Conditions: column, 3 m × 6.2 mm o.d., aluminum; column packing, 20% SF-96 on 60-80 mesh Chromosorb W; carrier, He, 60 ml/min; injector temp., 215°C; column temp., 50-250°C programmed at 4°C/min. Reprinted from Ref. 23 by courtesy of the American Oil Chemists' Society.

in a pyrolyzer (which was connected in series with a gas chromatograph) in an atmosphere of inert carrier gas (helium), and obtained a complicated chromatogram known as a "finger print" (see Fig. 4). Some of the peaks in the finger prints of linear alkylbenzenesulfonates (LAS) were identified as straight-chain 1-olefins of C_5-C_{10}, n-paraffins in the same molecular weight range as 1-olefins, and benzene, toluene, and xylene. The finger print of propylene-derived ABS (PPABS), however, did not show the regular spacing of linear-chain olefin and paraffin peaks as found for LAS, and the only peaks identified were those of benzene, toluene, and xylene. The branches of the alkyl group of PPABS seem to have been decomposed to lower molecular-weight hydrocarbons which gave rise to a crowd of peaks with short retention times in the finger print. Thus this technique can be used for qualitative and semi-quantitative analysis of hydrophobic-group distribution of ABS.

The complexity of the aforementioned pyrogram is obviously due to the random cracking of alkyl groups at the high decomposition temperature (650°C). To avoid such complexity, it is preferable to make a specific cleavage of the C—S bond rather than C—C bonds. In this regard, Lew's work [24a,b] is worth noting. He developed an on-line acid-pyrolysis gas chromatographic technique combining gas chromatography and the specific fission of C—S bonds by decomposing a mixture of a 1-5 mg of sample with 2-10 times its weight of P_2O_5 in a specially designed pyrolizer (see Fig. 5) at 400°C. The decomposition products were separated on a 67 m SF-96 capillary column. Each peak appearing on a chromatogram for LAS was identified with an authentic sample of AB, and showed good agreement with that of the precursor AB. On examination of the decomposition conditions, using sodium 2-phenyldodecanesulfonate as a standard sample, it was found that (a) pyrolysis at 650°C without acid, i.e., Liddicoet's conditions [23], lowered the yield of 2-phenyldodecane because of cracking of alkyl chains into low-molecular-weight pyrolyzates; (b) pyrolysis below 300°C slowed down the decomposition rate, resulting in broadening of the peak width; and (c) among the acids tested, either P_2O_5 or H_3PO_4 was preferred, sulfuric acid giving a poorer yield of 2-phenyldodecane due to oxidation. Figure 6 shows typical chromatograms of C_{11}-C_{14} LAS and its precursors; Table 2 lists the results for alkyl carbon number and isomer distributions calculated therefrom. This method, however, does not give quantitative results for PPABS because the resulting AB chromatogram is too complex to compare with authentic samples, while the mean position of the PPAB peak appears to depend on the degree of polymerization of propylene. A similar method confirming Lew's work was reported by Denig [25] who used a Currie point pyrolyzer [26] at 500°C, and identified the decomposition products by using GC-MS.

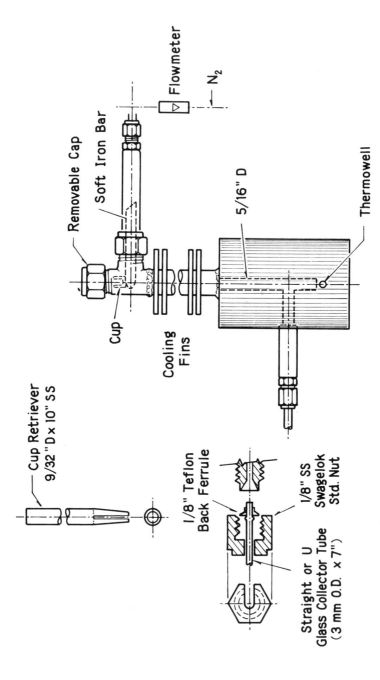

FIG. 5. Pyrolysis unit. All parts are stainless steel, except as noted. The pyrolysis block is wired (about 400W) and insulated. Reprinted from Ref. 24b by courtesy of the American Chemical Society.

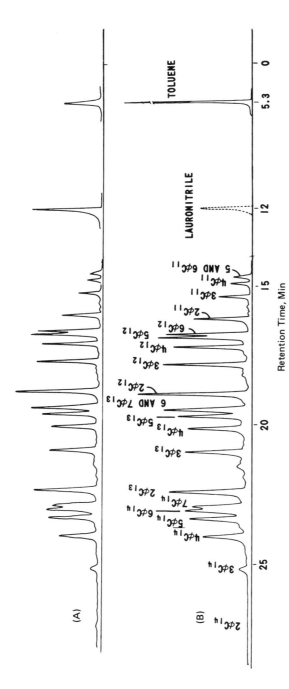

FIG. 6. P_2O_5-Pyrolysis of a mixture of sodium C_{11-14} LAS, lauric monoethanolamide (LEA), and sodium toluenesulfonate, 67 m × 0.5 mm i.d. stainless steel capillary column coated with SF-96 silicone oil, 120–220°C programmed at 16°C/min. A. P_2O_5-Pyrolysis of C_{11-14} LAS, LEA, and toluenesulfonate mixture. B. Chromatogram of C_{11-14} LA precursor and toluene mixture. Reprinted from Ref. 24a by courtesy of the American Oil Chemists' Society.

TABLE 2

P$_2$O$_5$ Pyrolysis of C$_{11}$-C$_{14}$ LAS in Detergent Formulations[a]

Alkylbenzene		C$_{11}$-C$_{14}$ Linear alkylate precursor			P$_2$O$_5$ Pyrolysis of C$_{11}$-C$_{14}$ LAS formulations Light duty[b]		Heavy duty[c]
Isomer distribution, wt %							
5- and 6-Phenyl	C$_{11}$	1.1	1.4	1.3	1.4	1.1	1.2
4-Phenyl	C$_{11}$	1.3	1.6	1.3	1.0	1.1	1.1
3-Phenyl	C$_{11}$	1.8	2.0	1.9	2.2	2.0	1.9
2-Phenyl	C$_{11}$	4.0	4.2	4.2	3.7	4.1	4.0
5- and 6-Phenyl	C$_{12}$	11.3	11.8	12.3	11.3	11.1	10.9
4-Phenyl	C$_{12}$	5.4	5.6	5.6	5.4	5.4	5.1
3-Phenyl	C$_{12}$	6.0	6.5	6.5	6.1	6.3	6.3
2-Phenyl	C$_{12}$	10.1	10.0	10.0	9.4	9.5	10.2
5-, 6-, and 7-Phenyl	C$_{13}$	14.0	13.6	14.4	14.2	13.6	13.7
4-Phenyl	C$_{13}$	5.0	5.5	5.6	5.4	5.5	5.3
3-Phenyl	C$_{13}$	5.5	5.9	6.2	6.0	6.4	6.2
2-Phenyl	C$_{13}$	9.6	8.7	8.6	7.8	9.1	9.5
5-, 6-, and 7-Phenyl	C$_{14}$	19.0	16.7	17.1	19.0	18.3	18.1
4-Phenyl	C$_{14}$	4.8	5.2	4.1	5.4	5.1	5.0
3-Phenyl	C$_{14}$	1.1	1.3	0.9	1.3	1.0	1.2
2-Phenyl	C$_{14}$	0.01	—	—	0.4	0.4	0.3
Side-chain carbon number distribution, wt %							
	C$_{11}$	8.2	9.2	8.7	8.3	8.3	8.2
	C$_{12}$	32.8	33.9	34.4	32.2	32.3	32.5
	C$_{13}$	34.1	33.7	34.8	33.4	34.6	34.7
	C$_{14}$	24.9	23.2	22.1	26.1	24.8	24.6
Average molecular weight		256.6	256.0	255.9	256.9	256.7	256.6

[a]Reprinted from Ref. 24a by courtesy of the American Oil Chemists' Society.
[b]18 parts C$_{11}$-C$_{14}$ LAS, 2 parts lauric monoethanolamide, and 3 parts p-toluene sulfonate.
[c]18% C$_{11}$-C$_{14}$ LAS, 2% lauric monoethanolamide, 3% p-toluene sulfonate, 45% TPP, 8% N-silicate, 1% CMC, 15% Na$_2$SO$_4$, and 8% water.

III. ALKYL SULFATES

A. Acid Hydrolysis

Alkyl sulfates, typified by sodium dodecyl sulfate (SDS), are the sodium salts of the esters of sulfuric acid and higher fatty alcohols, and have a C—O—S bond system between their hydrophobic and hydrophilic groups. This bond, in contrast to the stable C—S bond in ABS and alkylsulfonate, can be easily hydrolyzed in acid media to yield the original fatty alcohol as the hydrophobic oil. Investigation of the variation of the stability of SDS with reaction time, temperature and pH [27] showed that about 98% of SDS is hydrolyzed in 1 N HCl at 100°C in 7 hr, while ABS is quite stable under the same condition. This lability of alkyl sulfate is convenient for pretreatment for GC analysis and readily leads to a procedure in which the alkyl sulfate is hydrolyzed in acid media and the hydrophobic oil is extracted from the reaction solution with a suitable solvent, then gas chromatographed with or without further chemical treatment. In fact, fatty alcohols can be gas chromatographed either in unchanged form [28] or after converting to acetic acid esters [29]. In either case, it is not difficult to find suitable GC conditions for the separation of alkyl homologues and isomers of the fatty compounds. For the case of acetates, use of a stationary phase of the polyester type (e.g., DEGS) with temperature programming technique is recommended. Fig. 8.B shows an example for commercial SDS treated by this manner.

B. Acid Decomposition

The acid decomposition method is also applicable to the pretreatment of alkyl sulfates. Knight and House [6] applied their method (Sec. II.B) to alkyl sulfates and obtained the hydrophobic oil in good yield (72% theoretical for tallow alcohol sulfate, 80% for lauryl alcohol sulfate). Gas chromatographic analysis of the recovered oil indicated formation of olefins with carbon numbers corresponding to those of original fatty alcohols. Therefore, if a sample is a mixture of alkyl sulfate and ABS, the hydrophobic group distribution can be separately determined from the peaks due to each of the characteristic desulfonation products (olefins and alkylbenzenes).

C. On-line Pyrolysis

On-line pyrolysis gas chromatography of alkyl sulfate has been achieved in three different manners: (a) with no reagent, (b) with P_2O_5, and (c) with KOH, some of the pyrolyzates being common to all three techniques.

1. Pyrolysis Without Reagent

As mentioned above, olefinic pyrolyzates are obtained from acid decomposition of alkyl sulfate. This may be due to the dehydration of alcohol in the presence of conc. H_3PO_4, and, if so, pyrolysis without acid might be expected to leave the alcohol unchanged. The actual result was that at 650°C pyrolysis gas chromatography of commercial lauryl sulfate with no reagent gave major peaks due to C_{12} and C_{14} fatty alcohols and C_{12} and C_{14} 1-olefins, and minor peaks due to small amounts of 1-olefins from C_5-C_{11} [23]. The emergence of such major peaks indicates that this sample is actually a blend of lauryl and myristyl sulfates. The olefinic peaks seem due to partial thermal dehydration of alcohols, and the minor peaks due to further thermal degradation at the somewhat high temperature (650°C). The ratio of C_{12}-C_{14} alcohols present in the sample sulfate can be estimated with the intensity ratio of the peaks of C_{12}-C_{14} 1-olefins more reliably than those of C_{12}-C_{14} alcohols.

2. P_2O_5 Pyrolysis

On-line P_2O_5-pyrolysis gas chromatography gave peaks due only to a mixture of olefins, the formation of alkyl alcohols not being observed. This was also true for the case of P_2O_5-pyrolysis of fatty alcohol itself. Further separation of the olefinic peaks obtained from 1-nonyl sulfate showed that the olefins comprise 1-nonene and cis and trans isomers of possible internal nonenes (see Fig. 7). Therefore, if all the hydrophobic groups of alkyl sulfates are converted to the corresponding olefins in the same yield and the subsequent olefinic peaks having a certain carbon number do not overlap with those having other carbon numbers, it should be possible to determine the alkyl carbon number distribution of the original hydrophobic material by assuming that the olefins having equal carbon number show equal molar response. Though the detailed consideration on this point was not given by Lew [24a,b], the alkyl carbon number distribution of commercial tallow C_{14}-C_{18} alcohol sulfate (TAS) was successfully determined to ±5% accuracy (Table 3), which is satisfactory enough for most requirements.

This method also gives satisfactory results for blended surfactants in commercial formulations. Table 3 lists the results for the alkyl carbon number and isomer distributions of a blend of TAS and LAS. In this case, as well as in other blends, the pyrolysis products of a given surfactant do not change whether it is pyrolyzed alone or in a mixture of surfactants.

3. Alkali Fusion

On-line alkali-fusion gas chromatography of alkyl sulfate differs from the on-line techniques discussed above by producing small amounts of dialkyl

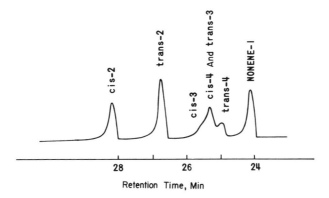

FIG. 7. Nonene isomers from P_2O_5-pyrolysis of 1-nonyl sulfate, 67 m × 0.5 mm i.d. stainless steel capillary column coated with SF-96 silicone oil, 60°C. Reprinted from Ref. 24a by courtesy of the American Oil Chemists' Society.

ether in addition to olefins and alcohols [30]. Nakagawa et al. [30] mixed and ground about 2 mg of alkyl sulfate mixture together with about 10 mg of KOH in a mortar and transferred the mixture to a pyrolyzer preheated to 400°C. The alkali fusion reaction took place instantaneously in the pyrolyzer and the reaction products were swept into PEG-containing column with a continuous flow of carrier gas.

The chromatogram thus obtained from a commercial SDS is shown in Fig. 8.A, where no peaks of alkyl alcohols and dialkyl ethers, however, are observed because of their low yields and long retention times under the conditions described. The alkyl carbon number distribution can be determined from the olefin peaks, each of which is due to internal and 1-olefins of equal carbon number. In Fig. 8.B a chromatogram of alkyl acetates derived from the acid hydrolysis product of the SDS is also shown for comparison. Satisfactory agreement is seen in these chromatograms and also in the numerical data for alkyl carbon number distribution and average molecular weight (see Table 4), which were calculated from these chromatograms.

Lew [24] and Nakagawa et al. [30] found that pyrolysis of alkyl sulfate without any reagent (P_2O_5 or KOH) produced small amounts of low-molecular-weight pyrolyzates. The same result was obtained by Liddicoet [23], suggesting merit in using chemical reagents for the selective cleavage of the bond between hydrophilic and hydrophobic groups.

TABLE 3

Pyrolysis of TAS and C_{11}–C_{14} LAS/TAS Formulation[a]

Analysis	Based on alcohol[b] from TAS	P_2O_5 Pyrolysis of TAS	Pyrolysis[c] of LAS/TAS formulation[d]	C_{11}–C_{14} Linear alkylate	P_2O_5 Pyrolysis of LAS/TAS formulation[d]
TAS alkyl chain carbon number distribution, wt %					
C_{14}	1.5	1.0	1.6		
C_{15}	0.8	0.6	0.7		
C_{16}	31.1	29.7	29.9		
C_{17}	3.4	3.9	3.6		
C_{18}	63.2	64.8	64.2		
TAS alkyl chain average molecular weight	242.6	243.3	224.9		
Alkylbenzene, isomer distribution, wt %					
5- and 6-Phenyl C_{11}				1.1	1.1
4-Phenyl C_{11}				1.3	1.2
3-Phenyl C_{11}				1.8	1.4
2-Phenyl C_{11}				4.0	4.5
5- and 6-Phenyl C_{12}				11.3	11.0
4-Phenyl C_{12}				5.4	6.4
3-Phenyl C_{12}				6.0	5.6
2-Phenyl C_{12}				10.1	11.1

5-, 6-, and 7-Phenyl	C_{13}	14.0	13.1
4-Phenyl	C_{13}	5.0	5.3
3-Phenyl	C_{13}	5.5	5.2
2-Phenyl	C_{13}	9.6	8.7
5-, 6-, and 7-Phenyl	C_{14}	19.0	18.7
4-Phenyl	C_{14}	4.8	4.9
3-Phenyl	C_{14}	1.1	1.3
2-Phenyl	C_{14}	0.01	0.5
Alkylbenzene side chain carbon number distribution, wt%			
	C_{11}	8.2	8.2
	C_{12}	32.8	34.1
	C_{13}	34.1	32.3
	C_{14}	24.9	25.4
Alkylbenzene average molecular weight		256.6	256.5

[a] Reprinted from Ref. 24a by courtesy of the American Oil Chemists' Society.

[b] Alcohol from dilute acid hydrolysis of TAS was analyzed in Aerograph 202 with 3.2 m, 62 mm 20% Carbowax 20M on Chromosorb W (HMDS); column temperature, 200°C; helium, 60 ml/min. Correction factors for alcohols were: C_{14}, 0.90; C_{16}, 1.00; C_{18}, 1.16; C_{15} and C_{17} assumed to be 1.00.

[c] Without acid.

[d] 10% C_{11-14} LAS, 5% TAS, 2% dodecanol-1, 45% TPP, 8% N-silicate, 1% CMC, 21% Na_2SO_4, and 8% water.

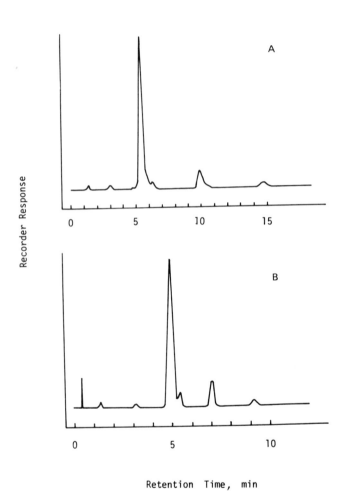

Retention Time, min

FIG. 8. Gas chromatogram of SDS. A. Olefinic products of alkali fusion of SDS; 20% PEG 20M on Chromosorb W(60-80 mesh), 1.5 m × 3 mm i.d., N$_2$ 40 ml/min, 120-240°C programmed at 6°C/min. B. Acetates from the same SDS; 10% diethyleneglycol succinate on Chromosorb W(60-80 mesh), 75 cm × 3 mm i.d., N$_2$ 40 ml/min, 150-220°C programmed at 10°C/min. Reprinted from Ref. 30 by courtesy of the Chemical Society of Japan.

TABLE 4

Alkyl Carbon Number Distribution of SDS[a]

Alkyl carbon number	Acetate		Alkali fusion	
	A	B[b]	A	B[c]
C_8	2.78	1.78	2.79	1.14
C_{10}	2.54	1.82	2.28	1.85
C_{12}	59.25	82.59	58.49	83.59
C_{14}	23.83	11.48	25.79	10.56
C_{16}	8.52	2.32	7.57	2.82
C_{18}	3.08	trace	3.08	trace
Average molecular weight	299.76	291.00	299.85	291.37
Average alkyl-chain carbon number	12.84	12.21	12.85	12.24

[a] Reprinted from Ref. 30 by courtesy of the Chemical Society of Japan.
[b] Corresponds to the chromatogram of Fig. 8.B.
[c] Corresponds to the chromatogram of Fig. 8.A.

IV. ALKYL- AND ALKENYLSULFONATES

α-Olefinsulfonate (AOS) has been developed as a biologically soft anionic surfactant and the chemistry of 1-olefins and their sulfonates (properties, syntheses, performance characteristics) has been intensively investigated. This surfactant is usually synthesized by sulfonation of 1-olefins obtained from either cracking of wax or polymerization of ethylene, and is a mixture of isomers and homologues having a wide distribution in carbon number (C_{10} to C_{20}). In the course of the synthetic processes, hydroxyalkanesulfonates (HAS) are formed as a major by-product [31]. Thus, the hydrophobic group of industrial AOS is usually a mixture of 1-olefins and hydroxyalkanes [32-34]. Hence, as a rule, AOS can be determined by measuring the degree of unsaturation and HAS by the hydroxyl values. However, these methods do not indicate the hydrophobic group distribution.

As mentioned in Sec. II, desulfonation of alkylarylsulfonates was successfully achieved by heating with phosphoric acid, but alkanesulfonates and α-olefinsulfonates were not desulfonated by this method; thus it was necessary to develop new techniques for the analysis of hydrophobic groups of AOS.

A. Pyrolysis Gas Chromatography of
   Sulfonyl Chlorides

In some cases, aliphatic, aromatic, and alkylarylsulfonates can be gas
chromatographed in the form of sulfonyl chlorides [18], but the technique
is limited to relatively short-chain sulfonyl chlorides because of the low
volatility and high degradability of long-chain ones. Such thermal instability
of long-chain sulfonyl chlorides prompted Nagai et al. [35] to subject AOS
to sulfonylization, followed by pyrolysis gas chromatography. To avoid the
complication of the sulfonylized products formed from alkenesulfonates in
AOS, the samples were first hydrogenated and then sulfonylized as shown
below:

$$RCH=CH(CH_2)_nSO_3Na \qquad\qquad\qquad RCH_2CH_2(CH_2)_nSO_3Na$$

$$\text{or} \qquad \xrightarrow[\text{Pt—H}_2]{\text{hydrogenation}} \qquad \text{or}$$

$$R'CH(CH_2)_mSO_3Na \qquad\qquad\qquad R'CH(CH_2)_mSO_3Na$$
$$\;|\qquad\qquad\qquad\qquad\qquad\qquad\qquad\;|$$
$$OH \qquad\qquad\qquad\qquad\qquad\qquad\qquad OH$$

$$RCH_2CH_2(CH_2)_nSO_2Cl$$

$$\xrightarrow[\text{DMF—SOCl}_2]{\text{sulfonylization}} \qquad\qquad \text{or} \qquad\qquad \longrightarrow \text{GC analysis}$$

$$R'CH(CH_2)_mSO_2Cl$$
$$\;|$$
$$Cl$$

The outline of their procedure follows:

Dissolve about 0.5-1.0 g of sample in 150 ml 50% ethanol, add a
similar weight of 5% palladium/charcoal and pass hydrogen through
the flask for 2 hr while stirring constantly and maintaining the temperature at 50°C in a water bath. Repeat the procedure twice in
the case of oleylsulfonate and check the completion of the reaction
by the disappearance of the 965 cm^{-1} band (C—H out-of-plane bend-
ing) from the infrared spectrum.

Mix the hydrogenated product with 20 ml thionyl chloride and 0.5 g
dimethyl formamide and reflux the resulting solution with agitation
under a nitrogen atmosphere at 80-85°C. Optimum reaction time
for C_{18} alkylsulfonate is 45 min; longer times produce oxidized
materials.

By this procedure, alkenylsulfonates and hydroxyalkylsulfonates are converted to alkyl sulfonyl chlorides and monochloroalkyl sulfonyl chlorides, respectively, and are then submitted to GC analysis on 3% SE-30 packed in 2.5 m × 4 mm i.d. glass column, at 200°C, carrier gas N_2 60 ml/min. The sample is injected directly onto the top of the glass column (see Fig. 9), whose temperature is controlled independently of the column temperature. The injected sulfonyl chlorides decompose instantaneously to give peaks of monochloroalkanes and 1-olefins from alkylsulfonyl chlorides and dichloro-alkanes, and monochloro-1-olefins from monochloroalkyl sulfonyl chlorides:

$$RSO_2Cl \longrightarrow RCl + R'CH{=}CH_2$$

$$\underset{\underset{Cl}{|}}{RSO_2Cl} \longrightarrow \underset{\underset{Cl}{|}}{R'Cl} + \underset{\underset{Cl}{|}}{R'CH}{=}CH_2$$

The structure of the products was confirmed by comparing GC-MS spectra and retention times of the peaks with those of authentic samples. The reproducibility of the decomposition products was examined with regard to the

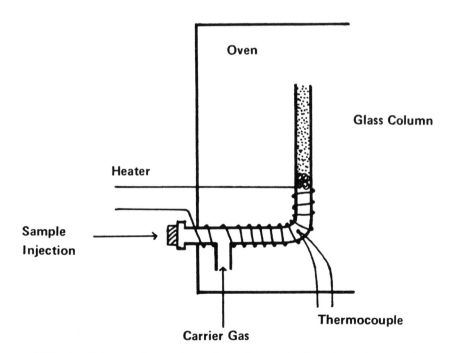

FIG. 9. Scheme of on-column injection port. Reprinted from Ref. 35 by courtesy of the American Oil Chemists' Society.

injection-port temperature and the sample amount injected onto the column.
Injection of 0.3-0.6 μl of 10% CCl_4 solution of the sample onto the column at
225°C gave the best results. Although it was difficult to eliminate olefin
formation completely, even under the conditions mentioned above, it was
found that this procedure permits quantitative determination with good reproducibility. After standardization with known amounts of C_{18} monochloroalkyl sulfonyl chloride, this method was applied to the analysis of the hydrophobic group of commercial AOS. Figure 10 shows the chromatogram thus
obtained for AOS derived from Chevron's C_{15}-C_{18} 1-olefin, which contains
disulfonate as is usually found in all such industrial products. The latter
undergoes the same chemical reactions and gives rise to an elevated baseline on the chromatogram. It is desirable to separate it by countercurrent
fractionation prior to this procedure if precise results are required. Table
5 gives alkyl carbon number distribution calculated from the chromatogram
of Figure 10 and indicates the good agreement with that of the original 1-
olefins.

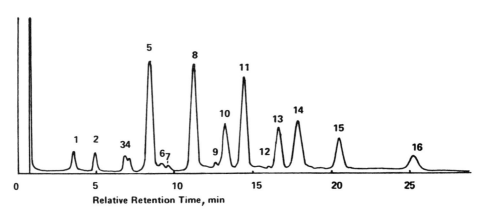

FIG. 10. Gas chromatogram of C_{15}-C_{18} sulfonyl chloride mixture by
programmed temperature GLC (170-200°C, 2°C/min). 1, C_{15}-α-olefin;
2, C_{16}-α-olfein; 3, C_{17}-α-olefin; 4, C_{15}-α-olefin chloride; 5, C_{15}-alkyl
chloride; 6, C_{18}-α-olefin; 7, C_{16}-α-olefin chloride; 8, C_{16}-alkyl chloride;
9, C_{17}-α-olefin chloride; 10, C_{15}-alkyl dichloride; 11, C_{17}-alkyl chloride;
12, C_{18}-α-olefin chloride; 13, C_{10}-alkyl dichloride; 14, C_{18}-alkyl chloride;
15, C_{17}-alkyl dichloride; 16, C_{18}-alkyl dichloride. Reprinted from Ref. 35
by courtesy of the American Oil Chemists' Society.

TABLE 5

GLC Analysis of AOS Derived from Chevron's C_{15}-C_{18} α-Olefin[a]

Carbon chain	RSO_2Cl	$R(Cl)SO_2Cl$	Total	Original α-olefin[b]
C_{15}	15.5	10.5	26.0	26.7
C_{16}	18.2	10.8	29.0	27.6
C_{17}	16.1	8.8	24.9	24.6
C_{18}	14.0	6.1	20.1	18.3
Total	63.8	36.2	100.0	97.2

[a] Reprinted from Ref. 35 by courtesy of the American Oil Chemists' Society.
[b] This sample contains 2.8% olefin other than C_{15}-C_{18}.

B. On-line Pyrolysis

On-line pyrolysis gas chromatography, either with or without reagents, gives only olefinic peaks from alkylsulfonates, and these peaks, similar to the ones of alkyl sulfates (Sec. III.C.2), comprise 1-olefins and cis and trans isomers of internal olefins. Therefore, for the determination of the alkyl carbon number distribution of alkylsulfonate, it is preferable to select conditions under which all olefinic peaks of a particular carbon number occur together rather than to attempt a full separation.

V. SOAP

Soap is a most popular detergent used since ancient days. Its hydrophobic portion, higher fatty acid, is usually supplied from natural sources, and mainly contains straight alkyl chains with even carbon numbers from C_8 to C_{20}. In acid media, soap is easily hydrolyzed to release free fatty acid which can be gas chromatographed rather simply after conversion to the methyl ester. Therefore, GC analysis of soap can be regarded as GC analysis of fatty acid itself.

A. Esterification

The fatty acid is usually esterified with methanol in the presence of acid catalysts such as hydrochloric acid [36], acetyl chloride [37], and boron trifluoride [38], or with diazomethane [39]. Among these reagents, BF_3-methanol is most suitable for the pretreatment for GC analysis and has been adopted by the American Oil Chemists' Society as a standard reagent for esterification [40, 41]. Use of $HClO_4$, instead of BF_3, as an acid catalyst has also been proposed [42], but in a heated system it is potentially hazardous [43]. Diazomethane is also widely used as the esterification reagent; whereas it reacts with free fatty acid to produce high yields of methyl ester and its volatility is convenient for removing the excess amount of the reagent after completion of the reaction, it requires careful handling because of its high toxicity.

B. Pyrolysis Gas Chromatography Combined
    with Ketone Formation

This technique was developed as a novel method for the analysis of long-chain fatty acid salts. Knowing that alkyl methyl ketone is formed by the thermal decarboxylation of a mixture of fatty acid salt and acetic acid salt [44], Nakagawa et al. [45] have investigated the applicability of this reaction in a pyrolyzer connected in series with a gas chromatograph under continuous stream of carrier gas, to the determination of the hydrophobic group distribution of soap. It was found that the yield of alkyl methyl ketone is affected by the type of metal cation, reaction temperature, and the mixing ratio of fatty acid and acetic acid salt. For instance, the yield of n-heptyl methyl ketone, which is formed from a mixture of caprylate and acetate salts (mixing ratio $C_8M/C_2M = 1/10$, reaction temperature 600°C), was found to increase in the order $Sr^{2+} > Cd^{2+} > Ba^{2+} > Ca^{2+} > Na^+$. The optimum reaction temperature proved to be 600°C, since reaction at 650°C caused the alkyl methyl ketones to decompose to lower-molecular-weight substances giving rise to minor peaks with short retention times, and reaction at 550°C slowed down the reaction rate to resulting in peak broadening and lowering in the yield of ketone.

Figure 11 illustrates these facts: the minor peaks appearing in the chromatogram on the left may be due to the low-molecular-weight substances, though it was not ascertained. It is obvious that this reaction produces dialkyl ketone and acetone at the same time. The yield ratio of such symmetric ketones and alkyl methyl ketone is closely related to the mixing ratio of fatty acid salt and acetic acid salt. In the case of a mixture of barium caprylate and acetate, the yield of n-heptyl methyl ketone at 600°C increased with weight ratio of C_2Ba/C_8Ba, and became constant (60%) in the region of 3:10. Thus, the established procedure for commercial soap is as follows:

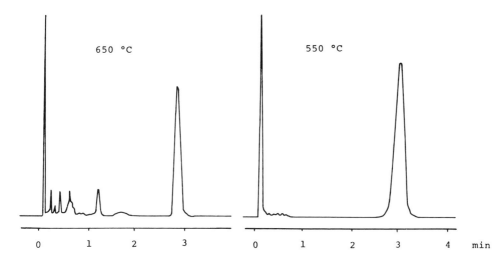

FIG. 11. Pyrolysis gas chromatogram of a mixture of barium caprylate and barium acetate at 650 and 550°C. Conditions: 20% Apiezon L, 75 cm, 140°C, 40 ml/min. Reprinted from Ref. 45 by courtesy of Preston Technical Abstracts Company.

Add an excess of hot aqueous barium hydroxide to the soap solution, remove the white precipitate thus formed by filtration and wash it with hot water until the filtrate is neutral. Dissolve the barium soap in a small volume of ethanol and reprecipitate it by addition of about 10 times the volume of water. Mix about 100 mg of the dried product thoroughly with 1 g of barium acetate in a mortar and transfer 3–5 mg of the mixture to the pyrolyzer preheated to 600°C.

The chromatogram obtained by this procedure is shown in Figure 12.A. The same soap was hydrolyzed and the resulting fatty acids were converted to methyl esters, whose gas chromatogram is shown in Fig. 12.B. In Fig. 12.A, it is seen that the peaks due to acetone and dialkyl ketones do not interfere with those of alkyl methyl ketones, but it is difficult to completely eliminate the minor peaks, possibly due to secondary degradation of ketones, which are observed in the short retention-time region. The numerical results calculated for these two chromatograms (see Table 6) indicate that both methods offer a quantitative determination of hydrophobic group distribution of soap.

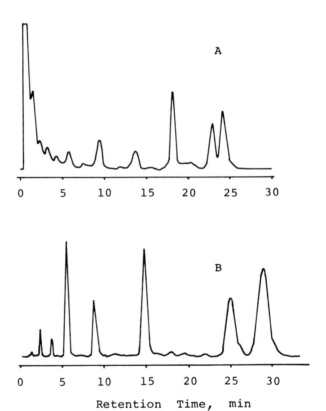

FIG. 12. A. Pyrolysis gas chromatogram of commercial soap. Conditions: column length, 3.0 m; column temp., 130-250°C; temp. programmed rate, 4°C/min; carrier gas flow rate, 40 ml/min. B. Gas chromatogram of methyl esters of fatty acids of commercial soap. Conditions: column length, 3.0 m; column temp., 200°C; carrier gas flow rate, 30 ml/min. Reprinted from Ref. 45 by courtesy of Preston Technical Abstracts Company.

TABLE 6

Alkyl Carbon Number Distribution of Soap[a]

Alkyl carbon number	Methyl ester, %	Alkyl methyl ketone, %
C_8	0.78	0.91
C_{10}	1.11	1.23
C_{12}	10.43	10.52
C_{14}	7.39	8.72
C_{16}	24.79	23.43
C_{18}	19.62	20.55
$C_{18}{=}$	35.88	34.64
Average carbon number	16.42	16.36
Average molecular weight[b]	283.11	282.38

[a]Reprinted from Ref. 45 by courtesy of the Journal of Chromatographic Science.
[b]As sodium salt.

VI. MISCELLANEOUS SURFACTANTS

There are many other anionic surfactants which occupy a relatively small share of surfactant production. Although these involve different kinds of hydrophilic groups from those dealt with in this chapter, their pyrolysis gas chromatographic behaviors can be forecast by taking into account the similarity of the bond combining the hydrophobic and hydrophilic groups. For instance, alkyl phosphate, involving C—O—P bond in its molecule, would be expected to show similar behavior to C—O—S bond in alkyl sulfate, and, in fact, alkali fusion gas chromatography of alkyl phosphates gives peaks due to 1-olefins and cis and trans isomers of internal olefins, together with smaller amounts of alcohol and dialkyl ether than those from alkyl sulfate. Such a similarity is also true in the relation between alkylphosphonate and alkylsulfonate, that is, alkali fusion produces an olefin mixture in both cases. Therefore, hydrophobic group analyses of alkyl phosphate and alkyl-phosphonate can be achieved by the same manner as alkyl sulfate and alkyl-sulfonate, respectively. In this connection, Lew's technique for P_2O_5

pyrolysis gas chromatography [24a,b] also seems applicable to these surfactants.

Alkyl and alkylphenol ethoxylated sulfates have both nonionic and anionic hydrophilic groups in the molecule, and may be analyzed by P_2O_5 pyrolysis gas chromatography [24a,b]. Their hydrophobic group distributions can be determined semiquantitatively by the peaks of hydrocarbons, e.g., tetrapropylene from tetrapropyleneethoxy sulfate and from tetrapropylenephenolethoxy sulfate.

REFERENCES

1. H. A. Gilman, Organic Chemistry, 2nd ed., Vol. 1, John Wiley and Sons, Inc., New York, 1943, p. 892.
2. W. J. Hickinbottom, Reactions of Organic Compounds, 3rd ed., Longman, Green, and Co., London, 1957, p. 564.
3. C. Friedel and J. M. Crafts, Compt. Rend. 109, 95 (1889).
4. V. Vesely and S. Stura, Coll. Czech. Chem. Commun. 6, 141 (1934).
5. V. Vesely and T. Stojanova, Decomposition of aromatic sulfonic acids by phosphoric acid. Coll. Czech. Chem. Commun. 9, 465 (1938).
6. J. D. Knight and R. House, Analysis of surfactant mixtures, I. J. Am. Oil Chemists' Soc. 36, 195 (1959).
7. S. Nishi, Simple gas chromatographic determination of aromatic sulfonates after desulfonation. Japan Analyst 14, 912 (1965).
8. E. A. Setzkorn and A. B. Carel, The analysis of alkylarylsulfonates by micro desulfonation and gas chromatography. J. Am. Oil Chemists' Soc. 40, 57 (1963).
9. E. R. Wright and A. L. Glass, Infrared analysis of detergents. Soap Chem. Specialities XLI, 59 (1965).
10. S. Lee and N. A. Puttnam, Gas chromatographic determination of chain-length distribution in fatty acid ethanolamides. J. Am. Oil Chemists' Soc. 42, 744 (1965).
11. S. Lee and N. A. Puttnam, Rapid desulfonation of alkylbenzenesulfonates. J. Am. Oil Chemists' Soc. 44, 158 (1967).
12. H. R. Henze and C. M. Blair, Chem. Eng. News 41, 130 (1963).
13. W. J. Carnes, Comparison of straight chain alkylbenzenes by gas chromatography. Anal. Chem. 36, 1197 (1964).
14. R. D. Swisher, The chemistry of surfactant biodegradation. J. Am. Oil Chemists' Soc. 40, 648 (1963).
15. E. Jungermann, G. A. Davis, E. C. Beck, and W. M. Linfield, Statistical approach to detergent evaluation. Correlation of performance data with gas chromatographic patterns of alkylbenzenes. J. Am. Oil Chemists' Soc. 39, 50 (1962).
16. F. Feigl and A. Lenzer, Microchim. Acta 1, 129 (1937).

17. S. Nishi, Simple identification of alkylarylsulfonates by alkali fusion combined with gas chromatography. Japan Analyst 14, 917 (1965).

18. J. J. Kirkland, Analysis of sulfonic acids and salts by gas chromatography of volatile derivatives. Anal. Chem. 32, 1388 (1960).

19. H. H. Bosshard, R. Mory, M. Schmidt, and H. Zollinger, A method for the catalyzed preparation of acyl and sulfonyl chlorides with thionyl chlorides. Helv. Chim. Acta 42, 1653 (1959) (in German).

20. W. Funasaka, T. Kojima, and Y. Toyota, Gas-liquid chromatographic analysis of naphthalenesulfonic acids. Japan Analyst 14, 815 (1965).

21. J. S. Parsons, Analysis of sulfonic acids by forming sulfonyl fluoride derivatives. J. Gas Chromatogr. 5, 254 (1967).

22. T. Kojima and H. Oka, Preprints 21th Annual Meeting Chem. Soc. Japan, No. 2, 1024 (1968).

23. T. H. Liddicoet and L. H. Smithson, Analysis of surfactants using pyrolysis-gas chromatography. J. Am. Oil Chemists' Soc. 42, 1097 (1965).

24a. H. Y. Lew, "Acid" pyrolysis-capillary chromatographic analysis of anionic and nonionic surfactants and some new developments in surfactant analysis. J. Am. Oil Chemists' Soc. 44, 359 (1967).

24b. H. Y. Lew, Preprints Am. Chem. Soc. Div. Petrol Chem. 16(3), B92 (1971).

25. R. Denig, Surfactant analysis by pyrolysis-gas chromatography. I. Nonionic and anionic surfactants. Tenside 10, 59 (1973) (in German).

26. W. Simon, P. Kriemler, J. A. Voellmin, and H. Steiner, Elucidation of the structure of organic compounds by thermal fragmentation. J. Gas Chromatogr. 5, 53 (1967).

27. M. Aoki and Y. Iwayama, Determination of ionic surface-active agents with dyes. IV. Applicability of fluorescein dyes as indicator, and stability of anionic detergents in solution. Yakugaku Zasshi (Tokyo) 80, 1749 (1960) (in Japanese).

28. W. E. Link, H. M. Hickman, and R. A. Morrissette, Gas-liquid chromatography of fatty derivatives. I. Qualitative analysis of n-alcohols. J. Am. Oil Chemists' Soc. 36, 20 (1959).

29. W. E. Link, H. M. Hickman, and R. A. Morrissette, Gas-liquid chromatography of fatty derivatives. II. Analysis of fatty alcohol mixtures by gas-liquid chromatography. J. Am. Oil Chemists' Soc. 36, 300 (1969).

30. T. Nakagawa, K. Miyajima, and T. Uno, Alkali fusion gas chromatography of alkyl sulfates. Bull. Chem. Soc. Japan 41, 2899 (1968).

31. T. H. Liddicoet, Alpha-olefins in the surfactant industry. J. Am. Oil Chemists' Soc. 40, 631 (1963).

32. D. M. Marquis, S. H. Sharman, R. House, and W. A. Sweeny, Alpha-olefin sulfonates from a commercial SO_3-air reaction. J. Am. Oil Chemists' Soc. 43, 607 (1966).

33. R. C. Odioso, Olefin sulfonates in detergents. Soap Chem. Speciali-
    ties 42, 47 (1967).
34. J. Rubinfeid and H. D. Cross, New trends in synthetic detergents.
    Soap Chem. Specialities 42, 41 (1967).
35. T. Nagai, S. Hashimoto, I. Yamane, and A. Mori, Gas chromato-
    graphic analysis for alpha-olefin sulfonate. J. Am. Oil Chemists'
    Soc. 47, 505 (1970).
36. W. Stoffel, F. Chu, and E. H. Ahrens, Jr., Analysis of long-chain
    fatty acids by gas-liquid chromatography. Anal. Chem. 31, 307 (1959).
37. R. K. Downey, Genetic control of fatty acid biosynthesis in rapeseed
    (brassica napus L). J. Am. Oil Chemists' Soc. 41, 475 (1964).
38. L. D. Metcalfe and A. A. Schmitz, The rapid preparation of fatty acid
    esters for gas chromatographic analysis. Anal. Chem. 33, 363 (1961).
39. H. Schlenk and J. L. Gellerman, Esterification of fatty acids with
    diazomethane on a small scale. Anal. Chem. 32, 1412 (1960).
40. Report of the Instrumental Techniques Committee, AOCS, 1966-1967,
    J. Am. Oil Chemists' Soc. 45, 103 (1968).
41. Report of the Instrumental Techniques Committee, AOCS, 1967-1968,
    J. Am. Oil Chemists' Soc. 46, 57 (1969).
42. P. J. Mavrikos and G. Eliopoulos, Preparation of methyl esters of
    long chain fatty acids. J. Am. Oil Chemists' Soc. 50, 174 (1973).
43. H. W. Wharton, Potential hazards of $HClO_4$ in heated systems where
    esters are involved. J. Am. Oil Chemists' Soc. 51, 35 (1974).
44. G. T. Morgan and E. Holmes, The higher methyl ketones. Chem. Ind.
    (London) 44, 108T (1925).
45. T. Nakagawa, K. Miyajima, and T. Uno, Pyrolysis-gas chromatog-
    raphy of long-chain fatty acid salts. J. Chromatogr. Sci. 8, 261 (1970).

Chapter 4

NUCLEAR MAGNETIC RESONANCE SPECTROMETRY OF ANIONIC SURFACTANTS

Hans König

Analytical Laboratories
Blendax-Werke R. Schneider GmbH and Co.
Mainz, Federal Republic of Germany

I. INTRODUCTION

A. General Aspects of NMR Spectroscopy

Nuclear magnetic resonance (NMR) spectrometry, especially proton magnetic resonance (PMR) spectrometry, is an excellent supplement to infrared (IR) and ultraviolet (UV) spectroscopy for the structure determination of organic compounds. Magnetic resonance signals are very sensitive to the intramolecular environment and both their locations and their structures (due to spin-spin coupling) yield useful information; PMR is particularly

Portions of this chapter are translated from Hans König, "NMR-Spectroskopie anionaktiver Tenside," Neuere Methoden zur Analyse von Tensiden, © 1971 Springer-Verlag, Berlin-Heidelberg-New York, by permission of the publisher.

suited to the qualitative and quantitative determination of hydrogen atoms in
different locations within a molecule.

Nuclear magnetic resonance spectra are usually obtained as a plot of
signal intensities versus change in strength of the applied magnetic field.
The applied field strength is expressed as the equivalent frequency units (in
Hz) divided by the frequency of the radio-frequency radiation utilized by the
spectrometer (in MHz) and is therefore obtained as a dimensionless quantity
in ppm. It is customary to add an internal standard whose signal is arbi-
trarily set at 0 ppm and the positions of all other signals (chemical shifts)
are measured relative to it in ppm or δ. Most chemical shifts from proton
resonance lie "downfield" of the commonest standard, tetramethylsilane
(TMS), and an alternative scale (not utilized in this chapter) is the tau system in which

$$\tau = 10 - \delta$$

In general terms, the magnetic-field strength at a nucleus is somewhat
less than that of the applied field because of the "shielding effect" of the
electrons around it, and the chemical shift shown by a particular hydrogen
nucleus depends upon the electron density around it. Thus adjacent electro-
negative groups or atoms (e.g., COO^-, N, S, O, multiple bonds, and aromatic ring systems) cause shifts to positions of greater (downfield) values
from TMS by reducing the electron density, and hence the shielding effect,
around the hydrogen atom.

The signal for a particular type of proton in a molecule is split into
two or more peaks by the presence of other types of hydrogen nuclei on
adjacent carbon atoms, a phenomenon known as "spin-spin coupling." Con-
sider the case of an ethyl group CH_3CH_2. In the methyl group, all three
hydrogen atoms are chemically equivalent, but there are four net spin-
energy levels for the CH_3 group, that is, (a) all three hydrogen atoms have
a spin of $+1/2$; (b) two hydrogen atoms have a spin of $+1/2$, and one hydrogen has a spin of $-1/2$; (c) one hydrogen atom has a spin of $+1/2$ and the
other two have spins of $-1/2$; (d) all three hydrogen atoms have a spin of
$-1/2$. The energy of the methylene hydrogen transition is thus split into
four discrete levels due to interaction with the four spin states of the methyl
group and appears in the spectrum as a quartet (of relative intensities
1:3:3:1), not as a single peak. For simple cases such as this, a particular
transition will yield n + 1 peaks, where n represents the number of other
types of hydrogen atoms which are magnetically equivalent and sited so that
such interactions can occur (usually on adjacent atoms). Thus spin-spin
coupling may permit determinations of the number of hydrogens on atoms
adjacent to a particular type of hydrogen atom.

The area under a peak is proportional to the number of the particular
type of hydrogen atoms present in the sample, and this permits the calculation of the relative abundance of various types of hydrogen nuclei in the

molecule. It is significant that this may be achieved without the necessity of pure standard substances or calibration curves.

Another important advantage of this technique is that the substances tested are not decomposed but may be recovered following evaporation of the solvent, provided that no standard has been added unless it, too, as is the case for TMS, is readily volatile.

B. Enhancement of Weak Signals

Spectra obtained from (a) dilute solutions of material, (b) minor components of a mixture, and (c) nuclei of low sensitivity such as ^{13}C are all expected to show weak signals and thus successful observations of these systems require some form of signal enhancement to increase the signal-to-noise ratio to acceptable levels. In principle it can be increased by extending the measuring period by sweeping through the spectrum over a period of several hours instead of a few minutes. However, this technique is of limited application because of low-frequency noise, changes in gain, and other sources of instability. A more practical method is to sweep through the spectrum rapidly many times in succession, commencing each new sweep at exactly the same point in the spectrum and then summing up the traces. This summation is conveniently achieved using a multi-channel pulse-height analyzer, which is essentially a small computer capable of recording the data of several hundred traces.

The recording of ^{13}C magnetic resonance spectra requires the accumulated data of perhaps several thousand scans, and although this is possible by conventional scanning methods, it is a tedious operation. An important alternate technique, known as Fourier transform NMR spectroscopy, utilizes the fact that the decay tail of the free induction signal following a resonance frequency pulse contains all the information of the slow-passage spectrum. Ernst and Anderson [1] have shown that such decay signals may be accumulated by a computer of average capacity to produce a high-resolution spectrum with a time saving on the order of 100- to 1000-fold.

None of the spectra described later in this chapter have been obtained in this manner and, at the time of writing, no Fourier transform NMR spectra of surfactants have been published. However, it is clear that this advancement will be of major importance and will increase rapidly in popularity. In particular it is expected to open new horizons in analytical chemistry by making ^{13}C NMR spectroscopy a more practical proposition.

C. Shift Reagents

A substantial improvement in the resolution of PMR spectra by addition of so-called "shift reagents" has been described by Hinckley [2] and Sanders

and Williams [3, 4]. Typical shift reagents are complexes of the rare
earths europium and praseodymium with such ligands as 2,2,6,6-tetra-
methyl-3,5-heptanedione or 1,1,1,2,2,3,3-heptafluor-7,7-dimethyl-4,6-
octanedione. They induce large changes in the chemical shifts of the PMR
spectra of compounds which possess functional groups with free electron
pairs capable of forming coordinate bonds with europium or praseodymium
ions. The improved resolution arises from the fact that the distribution of
electrons about those hydrogen atoms near to the coordination center undergoes a much greater change during coordination than does the electron distribution about the more remote protons. Thus the largest shifts occur for
protons closest to the functional group which alternates with the original
ligand.

The first application of shift reagents to structural studies of surfactants was by Stolzenberg et al. [5] to differentiate between the o- and p-alkyl
isomers of phenol polyglycol ether. The spectrum of the o-alkyl isomer
showed large changes in the position of some resonance signals following
interaction with a lanthanide complex, whereas the spectrum of the p-alkyl
isomer was but little influenced. The methylene protons of the ethylene
oxide groups of both isomers gave rise to two sets of triplets after com-
plexation.

II. NMR SPECTROSCOPY APPLIED
TO SURFACTANTS

A. Survey

With regard to the identification of surfactants, NMR spectroscopy yields
clearer information than other spectroscopic techniques concerning the
length and branching of alkyl chains, the lipophilic part of the molecule,
and the amount and location of double bonds. The degree of ethoxylation, if
any, can be determined quickly and readily by integration of the proton signals of the ethylene oxide (EO) groups. Aromatic compounds can be detected
without difficulty and, furthermore, the type of substitution can be specified
in many cases. However, it is necessary to appreciate that for commercial materials which often contain a number of isomeric and homologous
compounds, as well as by-products and soilings, the resultant NMR spectra,
in common with spectra obtained by other techniques, consist of an overlap
of the individual spectra of all compounds present and permit only average
structures to be determined.

Until recently, published NMR studies on surfactants have been confined mainly to nonionic surfactants. Thus Walz and Kirschnek [6] demon-
strated initially the determination of ethylene to propylene oxide ratio of
adducts of alkylphenols or fatty alcohols by this technique; Greff and Flanagan [7] described the PMR spectra of ethoxylated alkylphenols, fatty alcohols,

polypropylene oxides, ethoxylated mercaptans, fatty acid amides, and aliphatic amines; and Crutchfield, Irani, and Yoder [8] reported quantitative measurements on the PMR spectra of alkylaryl compounds and their ethylene oxide adducts.

Quantitative determination of the protons on the CH_2 group attached to a terminal OH group has been achieved by Page and Bresler [9] using pyridine as a solvent, since the CH_2OH resonance is thus shifted downfield away from the ethylene oxide envelope. However, the degree of separation is variable and it is frequently difficult to distinguish between some of the merging signals. Furthermore, interference by the OH resonance in the integration is still possible. Therefore Cross and Mackay [10] have suggested trimethylsilylation of the terminal OH group since, in PMR spectrum of such compounds treated with hexamethyldisilazane and trimethyl chlorosilane, the $OSi(CH_3)_3$ resonance occurs close to 0 ppm (TMS standard) and is thus far removed from most other proton resonances. Furthermore, a sharp, distinct internal standard (nine hydrogen atoms per molecule) is obtained enabling the number of hydrogen atoms responsible for the signals at 0.7-1.5 ppm and 3.0-3.9 ppm to be calculated and consequently the average values of the chain length and ethylene oxide content. This technique may be extended to the analysis of other hydroxylated compounds and, by analogy, to amines and thiols.

Nuclear magnetic resonance spectroscopy of ionic surfactants was first described by König [11, 12] (see below). Nagai et al. [13] determined the amount of 1-isomer in mixed alkenesulfonate by use of the fact that the olefinic protons of sodium 1-alkenesulfonate show chemical shifts at 6.1-6.7 ppm, whereas salts of the other alkenesulfonate give rise to signals at 5.4-5.9 ppm. Kuemmel and Liggett [14] subsequently confirmed that the 1-isomer gives rise to signals in the 6-7 ppm region in contrast to the 5-6 ppm region shown by isomers with nonterminal unsaturation. Because of low-intensity signals, computerized signal averaging was sometimes required.

B. Criteria for Obtaining Spectra

High-resolution NMR spectra may normally be obtained only from nonviscous liquids or solutions containing about 20% dry solid. To prepare such solutions (about 0.4 ml is normally required) a solvent containing no protons (e.g., carbon tetrachloride) is required, and at this stage it is appropriate to highlight one of the main problems in obtaining satisfactory spectra of ionic surfactants—that due to their limited solubility. Even in hydrophilic solvents such as deuterated water, most ionic surfactants are less than 10% soluble and if there is sufficient solubility above this level then the resultant solutions are highly viscous. Thus spectra are frequently of low intensity and poorly resolved. Hexa-deuterated dimethylsulfoxide (DMSO) proves to be a useful alternate solvent to deuterated water in such cases.

A wide range of anionic surfactants has been classified into groups (see Table 1). A commercial product typifying each of these groups has been selected for the purpose of recording the PMR spectrum and these are listed in Table 2 together with their chemical designation, trade name, and manufacturer.

TABLE 1

Classification of the Anionic Surfactants Examined

I. Salts of carboxylic acids

    A. Fatty acid soaps

    B. Aminocarboxylates

        1. Condensation products of amino carboxylic acids with fatty acid chlorides

            a. Fatty acid sarcosines

            b. Fatty acid protein condensates

        2. Condensation products of amino carboxylic acids with alkylsulfochlorides

II. Sulfated and sulfonated surfactants

    A. Sulfated surfactants

        1. Sulfated aliphatic alcohols

            a. Primary alkyl sulfates

            b. Secondary alkyl sulfates

        2. Sulfated nonionic surfactants

            a. Sulfated ethoxylated fatty alcohols

            b. Sulfated ethoxylated alkylphenols

        3. Sulfated oils and greases

    B. Sulfonated surfactants

        1. Alkanesulfonates

        2. Olefinsulfonates (alkenesulfonates and hydroxyalkanesulfonates)

        3. Alkylarylsulfonates

            a. Alkylbenzenesulfonates

                (1) Long-chain alkylbenzenesulfonates

(continued)

TABLE 1 (Cont.)

---

3. a. (2) Short-chain alkylbenzenesulfonates

   b. Alkylnaphthalenesulfonates

4. Sulfonates of nonionic surfactants

   a. Sulfonated ethoxylated alkylphenols

   b. α-Sulfonated fatty acid esters

   c. Sulfocarboxylic acid alkylol esters

5. Fatty acid derivatives of hydroxy- and amino-alkanesulfonic acids

   a. Fatty acid isethionates

   b. Fatty acid taurides

6. Sulfosuccinates

   a. Diester

   b. Monoester

III. Phosphated surfactants

   A. Phosphoric acid esters of fatty alcohols

   B. Phosphoric acid esters of ethoxylated fatty alcohols

Spectra were obtained from samples contained in 5 mm-diameter glass tubes using high-resolution 60 MHz spectrometers (Varian A 60 or JEOL JNM-MH-60 II) over a range of about 500 Hz for 1000 or 250 sec; integration times were 200 or 50 sec, respectively. TMS was used as a reference when using carbon tetrachloride or DMSO (6D) as a solvent, and the sodium salt of 2,2,3,3-tetradeutero-3(trimethylsilyl)propionic acid when using deuterated water; the value of signals from both of these reference materials has been assigned $\delta = 0$ ppm.

C. <u>Some Common Features of the Spectra of</u>
   <u>Anionic Surfactants</u>

Surfactants frequently contain long alkyl chains in which the forces acting upon most of the methylene protons are very similar; no splittings are observed in such cases. Multiplets of signals from isomeric mixtures overlap to yield the broad associating signals observed from many of the commercial products.

TABLE 2

Chemical Designations, Formulas, Trade Names and Manufacturers of the Examined Anionic Surfactants

Group Number	Chemical Designation	Formula	Trade Name	Manufacturer
I.A	sodium soap	$C_{15-17}H_{31-35}COONa$	———	Blendax, Werk Horb
I.B.1.a	sodium salt of coconut fatty acid sarcosine	$C_{11-17}H_{23-35}CO-N-CH_2-COONa$ $\mid$ CH_3	Medialan KA	Farbw. Hoechst
I.B.1.b	sodium salt of oleyl albuminate	$C_{17}H_{33}CONHR_1(CONHR_2)_x COONa$ x = 3-6	Lamepon A	Chem. Fabrik Grünau
	potassium salt of condensation products of protein hydrolyzates with plant fatty acids	$C_{11}H_{23}CONH[CH-R-CONH]_x CH-R-COOK$	Lamepon S	Chem. Fabrik Grünau
I.B.2	sodium salt of condensation products of protein hydrolyzates with alkylsulfochlorides	$R-SO_2-NH-R'(CONHR'')_n COONa$	Lamepon A 44	Chem. Fabrik Grünau
II.A.1.a	sodium lauryl sulfate	$C_{12}H_{25}O-SO_3Na$	Texapon K 12	Dehydag
	triethanolamine lauryl sulfate	$\left[C_{12}H_{25}OSO_3\right]^- \begin{bmatrix} & C_2H_4OH \\ HN-C_2H_4OH \\ & C_2H_4OH \end{bmatrix}^+$	Texapon TH	Dehydag
II.A.1.b	sec- sodium alkyl sulfate	$R-CH-CH_3$ $\mid$ OSO_3Na	Lensodel AB 6	Shell

Code	Description	Structure	Trade name	Manufacturer
II.A.2.a	sodium salt of ethoxylated lauryl sulfate	$C_{12}H_{25}(OC_2H_4)_{1-3}-OSO_3Na$	Texapon N 25	Dehydag
II.A.2.b	ammonium salt of ethoxylated sulfated alkylphenol	$R-\langle\text{benzene}\rangle-O(C_2H_4O)_x SO_3NH_4$	Fenopon CO 436	Gaf Corporation
II.A.3	sodium salt of sulfated ricinoleic acid	$CH_2-O-CO-(CH_2)_7CH=CH-CH_2-CH(CH_2)_5-CH_3$ with OSO_3Na; $CH-O-CO-(CH_2)_7CH=CH-CH_2-CH(OSO_3Na)(CH_2)_5-CH_3$; $CH_2-O-CO-(CH_2)_7CH=CH-CH_2-CH(CH_2)_5-CH_3$ with OSO_3Na	Avirol DKM	Böhme Fettchemie
II.B.1	sodium petroleum sulfonate	$R-SO_3Na$	Levapon T	Bayer
II.B.2	sodium α-olefin-sulfonate	$C_{12-14}H_{25-29}-CH=CH-SO_3Na$ and $C_{13-15}H_{27-31}-CH-SO_3Na$ with OH	Elfan OS 46	Hoesch-Chemie

(continued)

TABLE 2 (Cont.)

Group Number	Chemical Designation	Formula	Trade Name	Manufacturer
II.B.3.a(1)(a)	sodium dodecylbenzene-sulfonate	$C_{12}H_{25}$—⟨benzene⟩—SO_3Na	Marlon A 350	Chem. Werke Hüls
	triethanolamine dodecylbenzenesulfonate	$\left[C_{12}H_{25}-\langle\text{benzene}\rangle-SO_3\right]^- \left[HN\begin{smallmatrix}C_2H_4OH\\C_2H_4OH\\C_2H_4OH\end{smallmatrix}\right]^+$	Marlopon AT 50	Chem. Werke Hüls
II.B.3.a(1)(b)	sodium tetrapropylene-benzenesulfonate	$C_{12}H_{25}$—⟨benzene⟩—SO_3Na	Nansa HS 80/ scales	Marchon
II.B.3.a(2)	sodium toluene-sulfonate	H_3C—⟨benzene⟩—SO_3Na	Eltesol ST 90	Marchon
	sodium cumene-sulfonate	$\begin{smallmatrix}H_3C\\H_3C\end{smallmatrix}CH$—⟨benzene⟩—$SO_3Na$	Cumolsulfonat	Chem. Werke Hüls
II.B.3.b	sodium alkyl-naphthalene-sulfonate	⟨naphthalene⟩ with C_4H_9, C_4H_9, and SO_3Na	Nekal BX	BASF

Code	Name	Structural formula	Trade name	Manufacturer
II.B.4.a	sodium salt of ethoxylated octylphenol-sulfonate	C_8H_{17}—⬡—$(OC_2H_4)_x SO_3Na$	Triton X 200	Rohm & Haas
II.B.4.b	sodium α-sulfopalmitinic acid methyl ester	$CH_3-(CH_2)_{13}-\underset{\underset{SO_3Na}{\mid}}{CH}-COOCH_3$	——	Henkel & Cie.
II.B.4.c	sodium laurylsulfo-acetate	$C_{12}H_{25}-OOC-CH_2-SO_3Na$	Lathanol LAL	Stepan Chem. Comp.
II.B.5.a	sodium salt of coconut fatty acid isethionate	$C_{11-17}H_{23-35}COO-CH_2-CH_2-SO_3Na$	Hostapon KA	Farbw. Hoechst
II.B.5.b	sodium salt of coconut fatty acid tauride	$C_{11-17}H_{23-35}CO-NH-CH_2-CH_2-SO_3Na$	Hostapon KTW	Farbw. Hoechst
II.B.6.a	sodium di(2-ethylhexyl)-sulfosuccinate	$\begin{array}{l} CH_2-COO-C_8H_{17} \\ \mid \\ NaSO_3-CH-COO-C_8H_{17} \end{array}$	Manoxol OT	Hardman & Holden
II.B.6.b(1)	sodium salt of ethoxylated lauryl half ester of sulfosuccinic acid	$\begin{array}{l} CH_2-COONa \\ \mid \\ NaSO_3-CH-COO(C_2H_4O)_x C_{12}H_{25} \end{array}$	Steinapol SBF A 30	Rewo
II.B.6.b(2)	sodium salt of ethoxylated coconut alkylol-amide half ester of sulfosuccinic acid	$\begin{array}{l} CH_2COONa \\ \mid \\ NaSO_3-CH-CO(OC_2H_4)_x OR-NH_2 \end{array}$ in the relation 1:1 $+ NaSO_3-CH_2-CH-CO-NH-R-O(C_2H_4O)_x H$	Steinapol SBZ	Rewo

(continued)

TABLE 2 (Cont.)

Group Number	Chemical Designation	Formula	Trade Name	Manufacturer
II.B.6.b(3)	sodium salt of undecylenic acid monoethanolamide half ester of sulfosuccinic acid	$CH_2-COONa$ $\|$ $NaSO_3-CH-COO-CH_2-CH_2-NH-CO(CH_2)_8-CH=CH_2$	Steinazid SBU 185	Rewo
III.A	ethanolamine salt of sec-phosphoric acid ester of wax alcohols	$O=P \begin{array}{l} O-(CH_2)_x-CH_3 \\ O-CH_2)_x-CH_3 \\ O\left[H_3N(C_2H_4OH)\right] \end{array}$	Hostaphat KW 200	Farbw. Hoechst
III.B	sodium salt of sec phosphoric acid ester of oleyloctaglycol-ether	$O=P \begin{array}{l} O(C_2H_4O)_8C_{18}H_{35} \\ O(C_2H_4O)_8C_{18}H_{35} \\ ONa \end{array}$	Hostaphat KO 280	Farbw. Hoechst

The proton signal of a methyl group attached to methylene or methine appears at 0.8-0.9 ppm. Methylene protons give rise to signals at 1.3 ppm when situated adjacent to other methylene groups, at 2.0-2.2 ppm due to CH_2COOH, at 2.8 ppm due to CH_2-S-, at 4.1 ppm due to CH_2-O-S, and at 3.8 ppm when adjacent to a nitrogen atom. The CH_2-O group causes signals at 3.6-3.8 ppm in an ethoxylated compound and further downfield at 4.2 ppm in esters. Methylene protons sited between a nitrogen atom and a carboxylic group give a signal at 3.8 ppm.

Proton signals from alkylene groups with isolated double bonds appear at 5.3 ppm, whereas those from protons attached to carbon atoms in an aromatic ring structure appear at 7-8 ppm. In the case of para-substituted compounds, two symmetrical doublets from a AA'—XX' proton system result accompanied by more-or-less intensive bands or so-called "internal lines."

Commercially available deuterated solvents are frequently incompletely deuterated. Thus a quintet at 2.5 ppm, due to protons from the two methyl groups, is common in DMSO spectra. In wet samples an intense signal at 4.6-4.7 ppm appears, due to an exchange of the hydroxyl protons with the deuterium of the solvent, and is frequently accompanied by rotation side bands which are marked by linking lines in the recorded spectra.

A correlation table for the characteristic proton chemical shifts, already published by König [15], is given in a revised form in Table 3. It can be seen that the positions of the chemical shifts are generally constant within a narrow frequency range. Thus the nature of the functional group to which the protons in question belong may be readily determined.

Table 4 shows a comparison of the chemical shifts of similar atomic groups, revealing that they generally differ distinctly in their positions, a fact which simplifies identification considerably. The positions of the chemical shifts do not alter significantly in each of the solvents used under the applied conditions [11].

D. Description of Individual Spectra*

I.A. Fatty Acid Soaps

The electronic integration curve of the areas under the signals recorded in the spectrum of sodium soap indicates that there is a large proportion of shorter alkyl chains. Unsaturated fatty acids are recognized by the triplet at 5.3 ppm; the signal due to the protons of the CH_2- group adjacent to the double-bond appears at 1.6 ppm (Fig. 1).**

---

*Numbers and letters that identify compound groups refer to Table 2.
**Figures 1-30 appear on pages 161-190.

TABLE 3

Characteristic Positions of the Various Proton Signals of Anionic Surfactants[a]

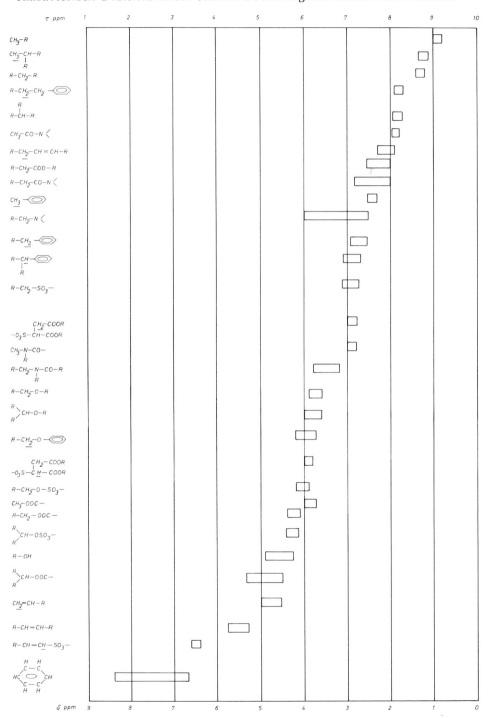

[a]In the formulas with several different protons, those protons related to the signals are underlined; R = H or alkyl.

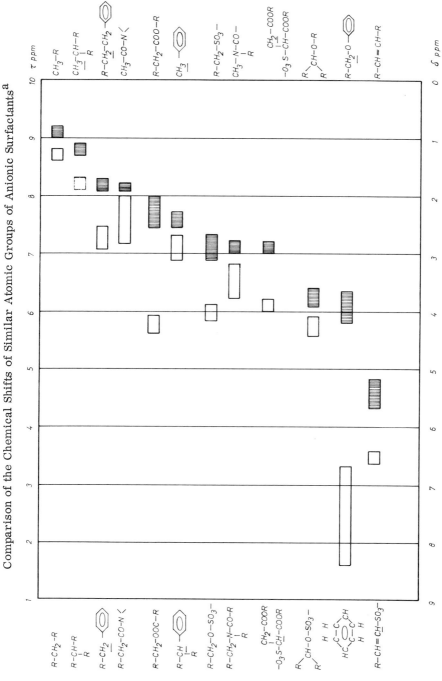

TABLE 4

Comparison of the Chemical Shifts of Similar Atomic Groups of Anionic Surfactants[a]

[a]In the formulas with several different protons, those protons related to the signals are underlined; R = H or alkyl.

I.B.1.a. Fatty Acid Sarcosines

Medialan KA shows a triplet at about 3 ppm due to the NCH_3- group, which is split by the adjacent methylene group. By the influence of the nitrogen atom the signal of this CH_2- group is shifted to 3.9 ppm (Fig. 2).

I.B.1.b. Fatty Acid Protein Condensates

The spectrum of Lamepon A shows, besides the proton signals due to the alkyl chain and the protons which are adjacent to the carboxylic group, proton signals due to the olefinic group at 5.3 ppm. The CH_2N- group can be detected by the signal at 3.7 ppm. Between 2.5 and 3.0 ppm and at 3.5 ppm other signals appear which cannot be readily assigned; probably they are due to protons of different amino acids from by-products (Fig. 3).

The spectrum of the Lamepon S is very similar to that of the Lamepon A, but it shows no signals due to double bonds (Fig. 4).

I.B.2. Fatty Acid Protein Condensates

The spectrum of Lamepon A 44 is generally similar to that of other Lamepons, apart from the presence of the signal due to the CH_2-S group at 2.8 ppm. Because the Lamepons are condensates of fatty acids and protein hydrolysates, no single structure adequately describes them and hence it is difficult to highlight any quantitative relationships (Fig. 5).

II.A.1.a. Primary Alkyl Sulfates

In the spectrum of Texapon K 12 the CH_2- group, which is adjacent to the sulfate group, is split into a triplet at about 4 ppm (Fig. 6).

The spectrum of Texapon TH reveals the triethanolammonium group very clearly with the two fully-symmetric signal groups due to the NCH_2- and the OCH_2- groups at 3.4 and 3.9 ppm; there the protons couple very strongly due to their flexibility, resulting in splittings of a higher order (Fig. 7).

II.A.1.b. Secondary Alkyl Sulfates

The integration of the proton signals of the Lensodel AB 6 shows that the product is not sulfated at the end of the alkyl chain (Fig. 8).

II.A.2.a. Sulfated Ethoxylated Fatty Alcohols

By the integration of the proton signals of the Texapon N 25 the average value of ethoxylation can be calculated; it is two moles of ethylene oxide per molecule (Fig. 9).

II.A.2.b. Sulfated Ethoxylated Alkylphenols

A branched alkyl chain, containing four methyl groups, can be detected in the spectrum of Fenopon CO 436. The substitution in the benzene ring has taken place in the para position; four protons, probably forming a AA'—XX' system, are detectable in the aromatic ring (Fig. 10).

II.A.3. Sulfated Oils and Greases

In the spectrum of the Avirol DKM besides the signals of the methyl and methylene protons, the signal due to the double bonds of the alkyl chains at 5.4 ppm is recognized, overlapping with the proton signal due to the CHO— group of the glycerol ester at 5.2 ppm. The CH_2— group adjacent to the carboxylic group appears at 2.2 ppm together with the CH_2— group adjacent to the methine group. The proton signals due to the CHOS— group and the CH_2O— group of the glycerol overlap at 4.2 ppm. A signal due to the methine group of the secondary alcohol at 3.6 ppm indicates the presence of diester in addition to triester (Fig. 11).

II.B.1. Alkanesulfonates

The signal-strength ratio of the methyl protons to the other protons in the spectrum of Levapon T indicates that the product is only slightly sulfonated at the end of the alkyl chain (Fig. 12).

II.B.2. α-Olefinsulfonates

In the spectrum of Elfan OS 46, signals appear from the methyl group (0.9 ppm); from methylene protons adjacent to other methylene groups in the alkyl chain (1.3 ppm), adjacent to the olefinic methine group, and probably also in the 2-position to the sulfonate group (1.8-2.2 ppm), and attached to the sulfur atom (3.0 ppm); and from methine protons adjacent to the hydroxyl group (3.6 ppm), adjacent to a nonterminal double bond (5.7 ppm) and in the 1-position to the sulfonate group (6.5 ppm). Integration of the different proton signals indicates that only a small proportion of the olefinic double bonds and hydroxyl groups are in the 1-position. A quantitative interpretation of the chemical shifts can only give average values, because there are too many isomers of the alkene- and hydroxyalkane sulfonates (Fig. 13).

II.B.3.a(1)(a). Long-chain Alkylbenzenesulfonates

The aromatic character of Marlon A 350 can be identified by the two symmetric doublets between 7 and 8 ppm, which indicate aromatic hydrogen atoms in a para-substituted benzene ring. The broad overlapping signals of the methyl and methylene protons reveal that there is a mixture of various isomers, i.e., the benzene ring is not situated at the end of the alkyl chain but that it may be attached at any methylene group of the alkyl chain (Fig. 14).

In Marlopon AT 50, in addition to para-substitution, the triethanolammonium structure can be seen very clearly from the two symmetric signal groups between 3 and 4 ppm (Fig. 15).

II.B.3.a(1)(b). Branched Long-chain Alkylbenzenesulfonates

The relatively large number of protons in the 0.9 ppm region of the spectrum
of Nansa HS 80 scales indicates that the alkyl group is highly branched (Fig. 16).

II.B.3.a(2). Short-chain Alkylbenzenesulfonates

In the spectrum of Eltesol ST 90 the signal due to the methyl protons, which
are in para-substitution to the sulfonate group of the aromatic system, is
shifted to 2.4 ppm. Two signals appear at 1.3 ppm and a triplet at 2.6 ppm,
indicating a by-product, which, according to the chemical shifts, could possibly be an isopropyl compound (Fig. 17).
    In the spectrum of Cumolsulfonate the para-substitution can be distin-
guished by the two doublets between 7 and 8 ppm. At about 2.9 ppm a sep-
tet of the proton due to the CH— group adjacent to the aromatic ring is
formed; splitting, however, is rather indistinct. The signal due to the two
methyl groups appears under the doublet at 1.2 ppm (Fig. 18).

II.B.3.b. Alkylnaphthalenesulfonates

The spectrum of Nekal BX is very unstructured and shows no splitting due
to spin-spin coupling. Between 7.2 and 8.5 ppm rather symmetric signals
appear due to protons of the aromatic system. At 2.6 ppm the signal due
to the CH_2— protons adjacent to the aromatic rings arises. The broad associating signals of the alkyl chains, which appear between 0.6 and 1.8 ppm
indicate the presence of an isomeric mixture (Fig. 19).

II.B.4.a. Ethoxylated Alkylphenolsulfonates

The spectrum of Triton X-200 shows numerous signals including doublets
due to a para-substituted aromatic ring (6.7 and 7.4 ppm), that of a CH_2O—
group adjacent to the ring (4 ppm), and of the protons of the ether groups at
3.6 ppm; the CH_2— group in the para-position forms a sharp signal at 3.3
ppm and the adjacent CH— group, also influenced by the benzene ring, is
indicated at 1.7 ppm; the triplet at 2.8 ppm is probably due to the CH_2—
group adjacent to SO_3—. The signal at 2.5 ppm is due to the solvent, dimethylsulfoxide, which is probably not fully deuterated. Integration of the
proton signals of the alkyl chain indicate that it is branched; the strong,
well-separated signals from CH_3— and CH_2— protons are in contrast to the
broad, associating signals shown by other types of alkyl chain (Fig. 20).

II.B.4.b. α-Sulfonated Fatty Acid Esters

In the spectrum of α-sulfopalmitinic acid methyl ester sodium salt signals
appear due to CH_3—, CH_2—, and CH_2— adjacent to CH—, in the alkyl chain

at 0.9, 1.3, and 2 ppm, respectively. The signal from the methine group between the carboxylate and sulfonate groups overlaps with that of the CH_3O- group at 3.8 ppm; if deuterated DMSO is used as a solvent the combined signal splits into two single peaks due to CH_3O- (3.5 ppm) and $CH-$ (3.7 ppm) (Fig. 21).

II.B.4.c. Alkylsulfocarboxylic Acid Esters

As in the previous spectrum, Lathanol LAL shows signals due to CH_3-, CH_2-, and the methine group between the carboxylate and sulfonate groups at 0.9, 1.3, and 3.5 ppm, respectively. The CH_2- group attached to the oxygen atom of the fatty alcohol group gives rise to the peak at 4.2 ppm. Water is the probable cause of the 3.1 ppm signal (Fig. 22).

II.B.5.a. Fatty Acid Isethionates

Signals due to methylene protons in Hostapon KA are well-separated, i.e., at 2.2 ppm for CH_2CO-, at 2.8 ppm for CH_2S-, and at 4.1 ppm for CH_2OOC-, the latter two groups yielding triplets. Alkyl-chain protons cause the signals at 0.8 and 1.2 ppm (Fig. 23).

II.B.5.b. Fatty Acid Taurides

The spectrum of Hostapan KTW neu required a large amplification to overcome the poor solubility of the material; hence the low signal-to-noise ratio. The signal at 3.7 ppm is a foreign one, possibly resulting from traces of water (Fig. 24).

II.B.6.a. Diester of Sulfosuccinic Acid

Signals due to the $CH-CH_2$ grouping in the succinyl portion of Manoxol OT appear at 3.2 and 4 ppm, the latter overlapping with the signal from CH_2OOC- of the ester. Integration of the alkyl CH_3- and CH_2- groups indicate that branched (ethylhexyl) chains are present.

II.B.6.b. Half-esters of Sulfosuccinic Acid

Chemical shifts due to the CH_2CH- of the sulfosuccinyl group are evident in all three classes below, but are at least partly obscured by overlap with other peaks; methylene protons from ethylene oxide yield their characteristic peaks at 3.6 ppm.

II.B.6.b(1) Sodium Salt of Ethoxylated Lauryl Half-ester of Sulfosuccinic Acid

In the Steinapol SBF A 30 spectrum, alkyl-chain protons typically give rise to signals at 0.8 and 1.3 ppm. The CH_2O- group peak overlaps with that of the sulfosuccinyl CH— group at 4 ppm. Peaks at 2.1 and 3.0 ppm are believed due to amide groups in impurities (Fig. 26).

II.B.6.b(2). Sodium Salt of Ethoxylated Coconut Alkylolamide Half-ester of Sulfosuccinic Acid

A high degree of peak overlap exists in the spectrum of Steinapol SBZ. Sulfosuccinyl CH_2- overlaps with CH_2- (adjacent to NH_2) at 2.7-3.0 ppm; amide CH_2- probably occurs in the 3.4-3.7 ppm region together with ethylene oxide; and ester CH_2O- overlaps with the sulfosuccinyl CH— group. Two peaks at 0.9 and 2.1 ppm indicate the presence of CH_3- and CH_2CO- groups which cannot be attributed to the main components and probably arise from the presence of a fatty acid alkanolamide (Fig. 27).

II.B.6.b(3). Sodium Salt of Undecylenic Acid Monoethanolamide Half-ester of Sulfosuccinic Acid

In addition to the 1.3 ppm signal of alkyl-chain CH_2- protons, the spectrum of Steinazid SBU 185 shows peaks due to vinyl (5 ppm), methine (5.6 ppm), and CH_2- groups adjacent to methine (2.1 ppm), the latter overlapping with signals from CH_2- adjacent to the carboxylate ester group and from the CH_2CO- group of undecylenic monoethanolamide, which is known to be present as an impurity. The sulfosuccinyl CH_2CH- signals appear among the multiplet at 3 ppm and partly overlap with the CH_2O- peak of the monoethanolamide at 4.2 ppm. Protons from CH_2- adjacent to nitrogen can be distinguished at 3.5 ppm (Fig. 28).

III.A. Phosphoric Acid Esters of Fatty Alcohols

In the spectrum of the Hostaphat KW 200 the signal due to the proton of the CH_2OP- group appears at about 3.8 ppm (Fig. 29).

III.B. Phosphoric Acid Esters of Ethoxylated Fatty Alcohols

The spectrum of the Hostaphat KO 280 shows, in addition to the triplet due to the double bonds at 5.3 ppm, a signal at about 2 ppm due to the protons of the CH_2- group adjacent to the methine groups. The ethylene oxide groups as usual can be detected by the signal at 3.6 ppm (Fig. 30).

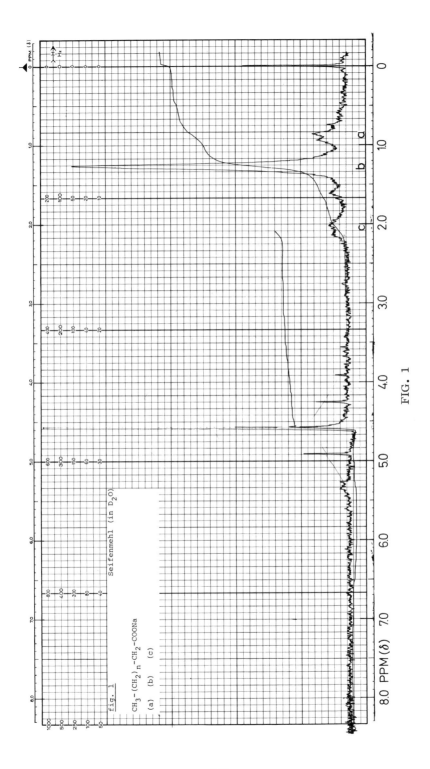

FIG. 1

Seifenmehl (in D$_2$O)

fig. 1

CH$_3$-(CH$_2$)$_n$-CH$_2$-COONa

(a)   (b)   (c)

161

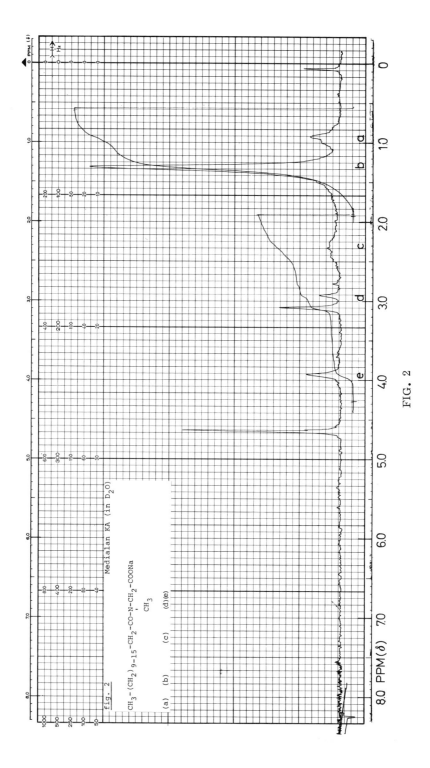

fig. 2

Medialan KA (in D₂O)

$CH_3-(CH_2)_{9-15}-CH_2-CO-N-CH_2-COONa$
$\qquad\qquad\qquad\qquad\quad |$
$\qquad\qquad\qquad\qquad\ CH_3$

(a) (b)　(c)　(d)(e)

FIG. 2

162

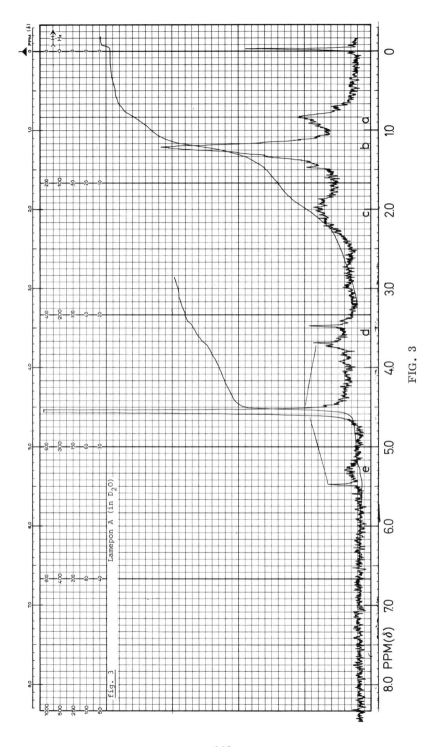

Lamepon A (in D₂O)

fig. 3

FIG. 3

163

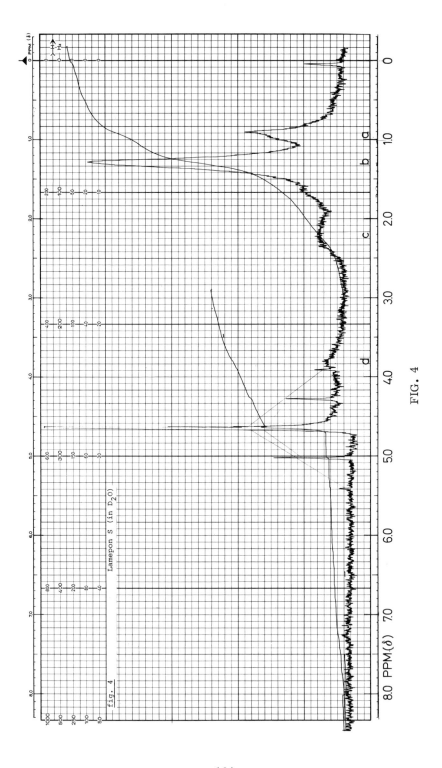

Lamepon S (in D₂O)

fig. 4

FIG. 4

164

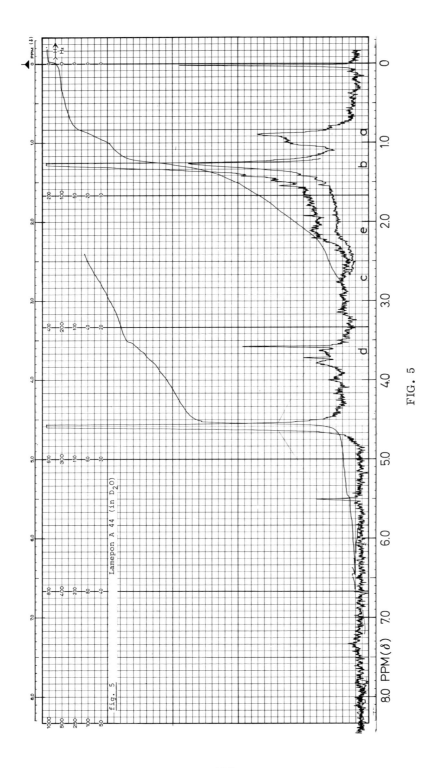

FIG. 5

165

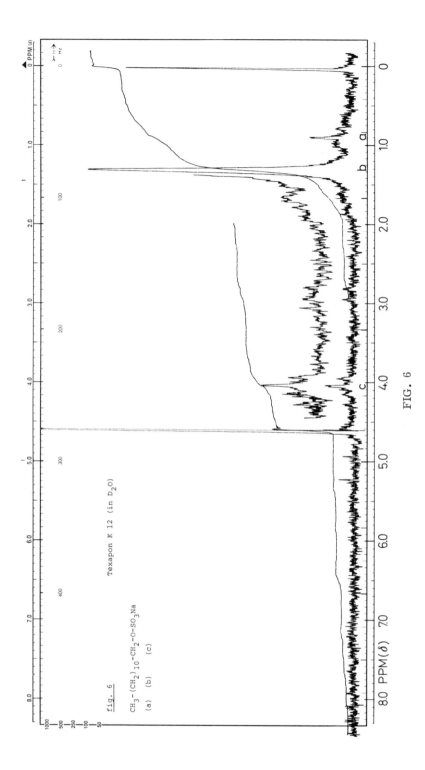

Texapon K 12 (in D$_2$O)

fig. 6

CH$_3$-(CH$_2$)$_{10}$-CH$_2$-O-O-SO$_3$Na
(a)   (b)   (c)

FIG. 6

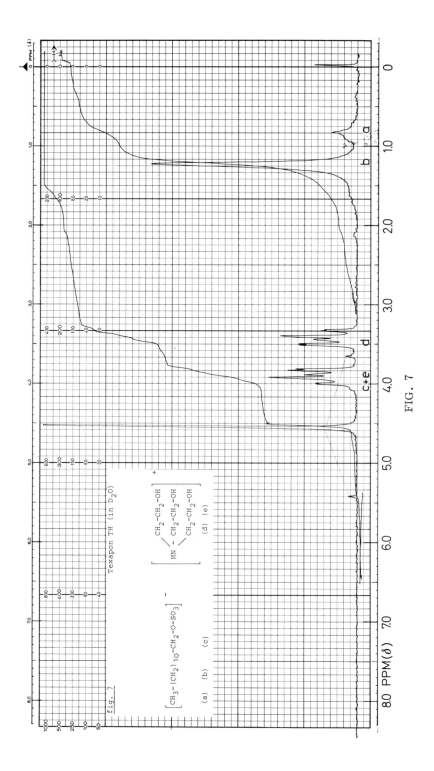

FIG. 7

167

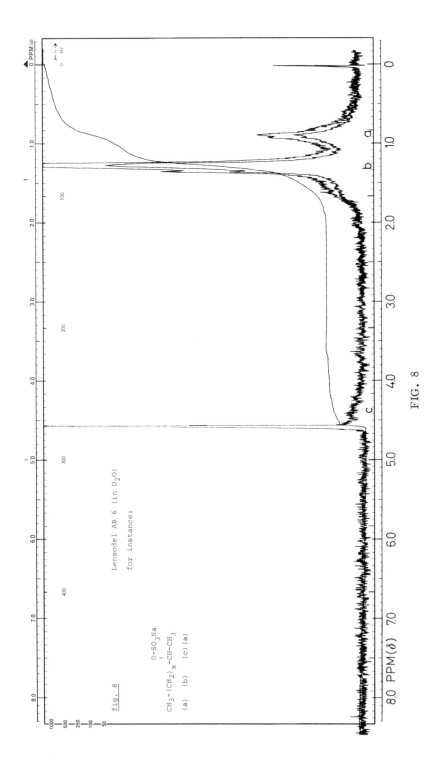

FIG. 8

Lensodel AB 6 (in D_2O)

for instance:

fig. 8

$$CH_3-(CH_2)_x-CH-CH_3$$
$$\qquad\qquad\quad |$$
$$\qquad\qquad O-SO_3Na$$

(a) (b)   (c)(a)

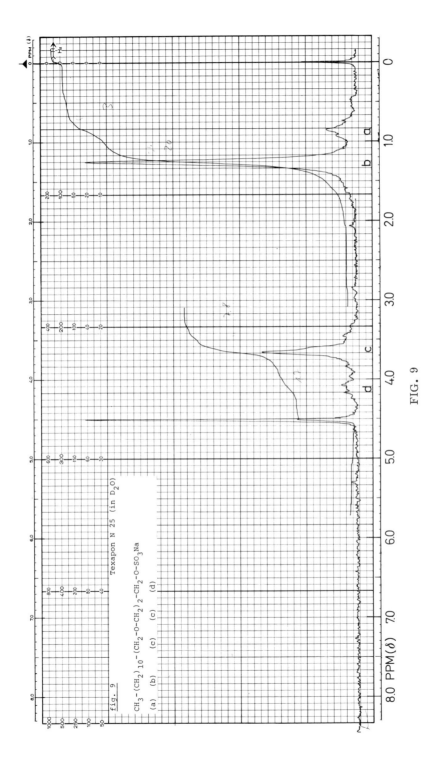

Texapon N 25 (in D$_2$O)

fig. 9

$CH_3-(CH_2)_{10}-(CH_2-O-CH_2)_2-CH_2-O-SO_3Na$

(a) (b) (c) (c) (c) (d)

FIG. 9

169

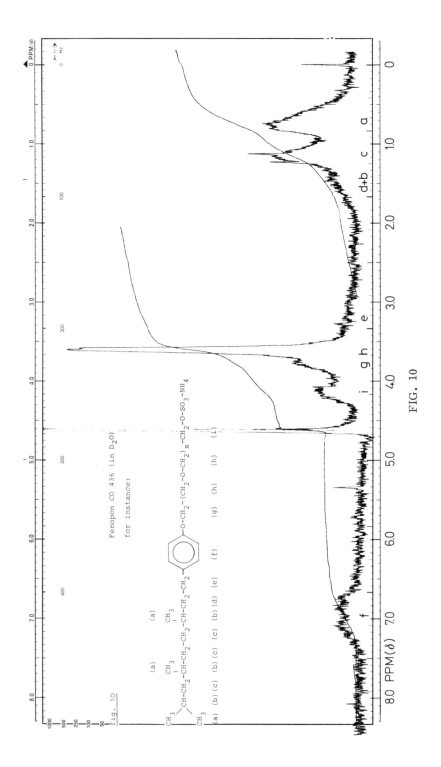

Fenopon CO 436 (in D₂O);

for instance:

$CH_3-CH-CH_2-CH-CH_2-CH-CH_2-CH_2-CH_2$

with CH_3 groups

$-O-CH_2-(CH_2-O-CH_2)_x-CH_2-O-SO_3-NH_4$

(a)(b)(c) (c) (b)(d) (e)   (f)   (g) (h)   (h)   (i)

FIG. 10

170

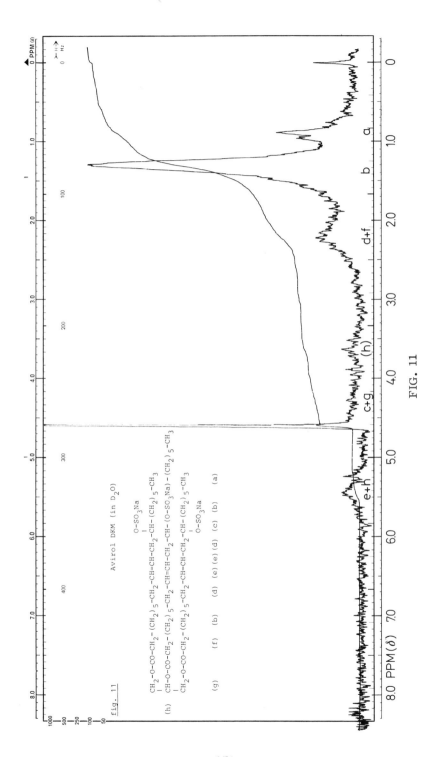

FIG. 11

Avirol DKM (in D$_2$O)

$$
\begin{array}{l}
\text{(h)} \quad \underset{\substack{\text{(h)}}}{\text{CH}_2}\text{-O-CO-CH}_2\text{-(CH}_2)_5\text{-CH}_2\text{-CH=CH-CH}_2\text{-CH-CH}_2\text{-CH-(CH}_2)_5\text{-CH}_3 \\
\qquad\qquad\qquad\qquad\qquad\qquad\qquad\qquad\qquad\overset{\displaystyle|}{\text{O-SO}_3\text{Na}} \\
\quad \text{CH-O-CO-CH}_2\text{-(CH}_2)_5\text{-CH}_2\text{-CH=CH-CH}_2\text{-CH-(O-SO}_3\text{Na)-(CH}_2)_5\text{-CH}_3 \\
\quad \text{CH}_2\text{-O-CO-CH}_2\text{-(CH}_2)_5\text{-CH}_2\text{-CH=CH-CH}_2\text{-CH-CH}_2\text{-CH-(CH}_2)_5\text{-CH}_3 \\
\qquad\qquad\qquad\qquad\qquad\qquad\qquad\qquad\qquad\overset{\displaystyle|}{\text{O-SO}_3\text{Na}}
\end{array}
$$

(g)   (f)   (b)   (d) (e) (e) (d) (c) (b)   (a)

fig. 11

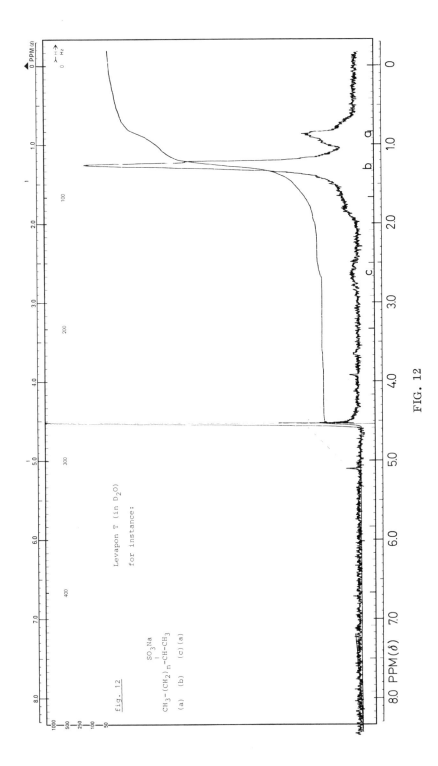

FIG. 12

172

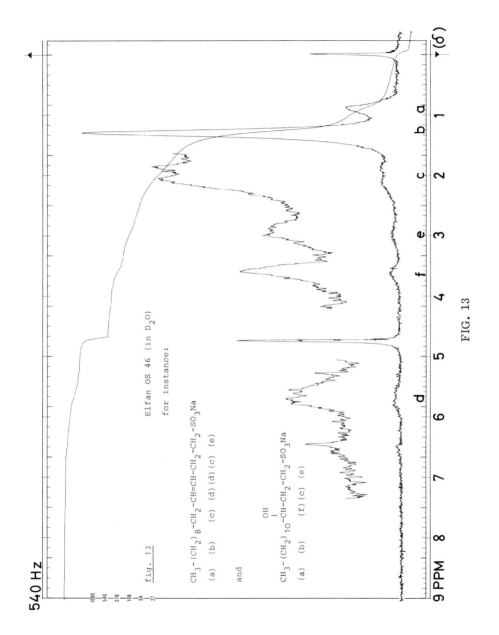

540 Hz

1080
540
270
108
54
27

Elfan OS 46 (in D$_2$O)

for instance:

fig. 13

$CH_3-(CH_2)_8-CH_2-CH=CH-CH_2-CH_2-SO_3Na$
(a)   (b)   (c) (d) (d) (c)   (e)

and

$CH_3-(CH_2)_{10}-\overset{\overset{\displaystyle OH}{|}}{CH}-CH_2-CH_2-SO_3Na$
(a)   (b)   (f) (c) (e)

FIG. 13

9 PPM   8   7   6   5   4   3   2   1   →(δ)

d   f e   c b a

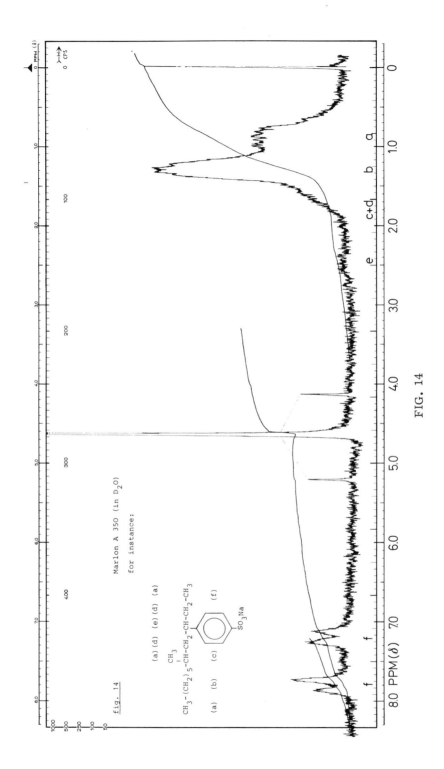

FIG. 14

174

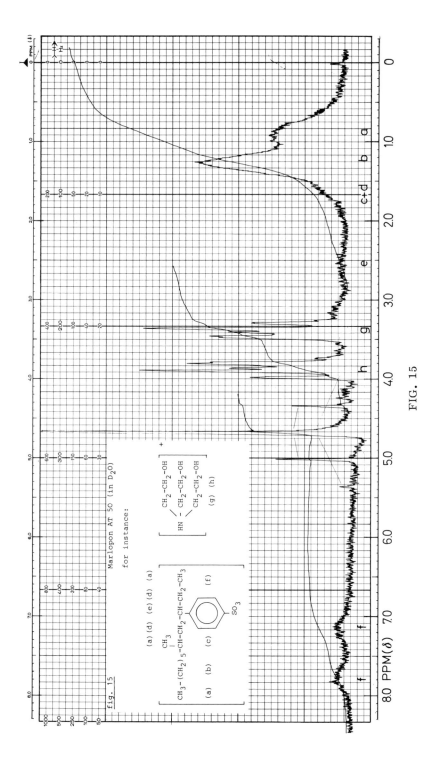

FIG. 15

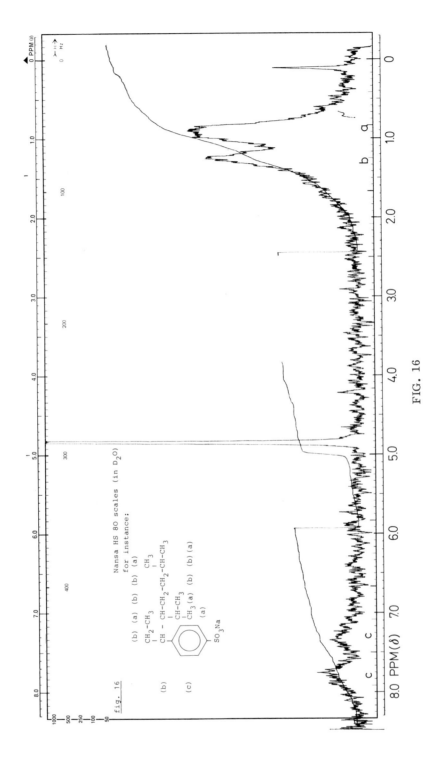

Nansa HS 80 scales (in D$_2$O)
for instance:

(b)   (a) (b) (b) (a)   CH$_3$
CH$_2$–CH$_3$ |
(b)   CH – CH–CH$_2$–CH$_2$–CH$_2$–CH–CH$_3$
          |
          CH–CH$_3$
(c)   CH$_3$(a) (b) (b) (a)
          (a)

          SO$_3$Na

f i g . 16

FIG. 16

176

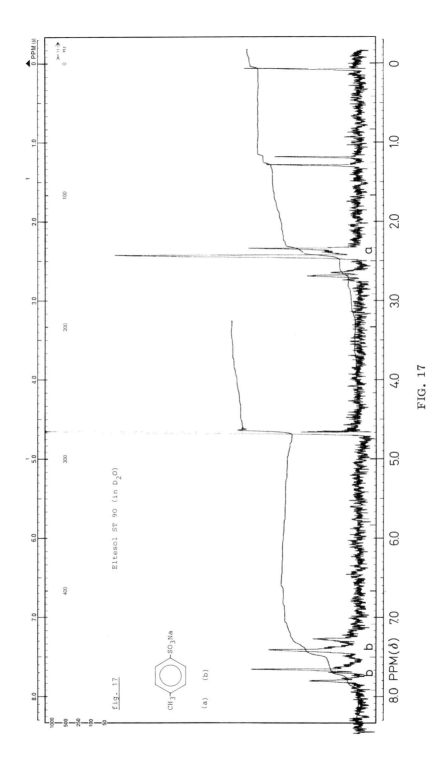

Eltesol ST 90 (in D₂O)

fig. 17

FIG. 17

177

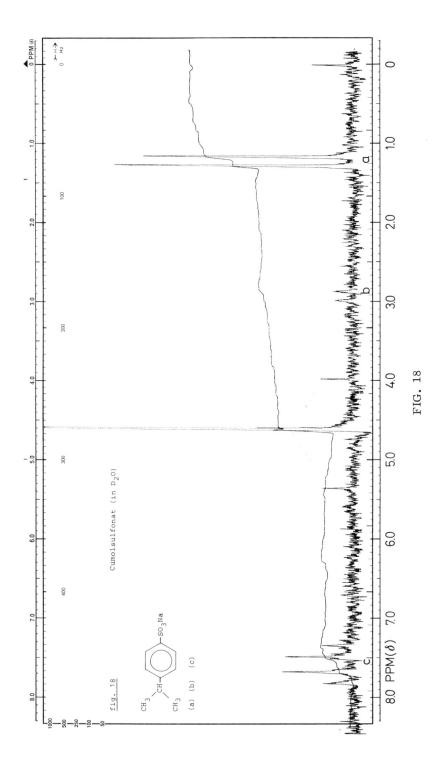

Cumolsulfonat (in D$_2$O)

fig. 18

FIG. 18

178

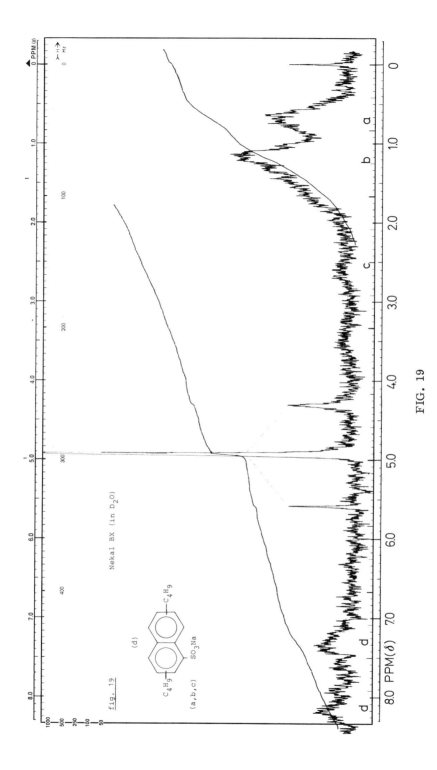

Nekal BX (in D₂O)

fig. 19

(d)

(a,b,c)

FIG. 19

179

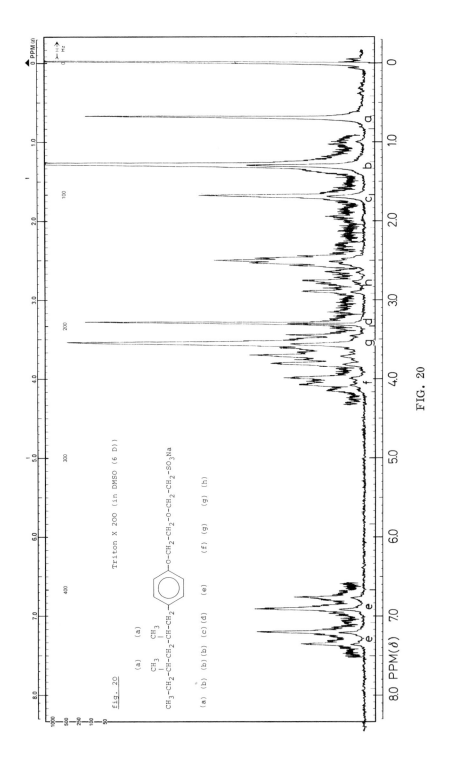

FIG. 20

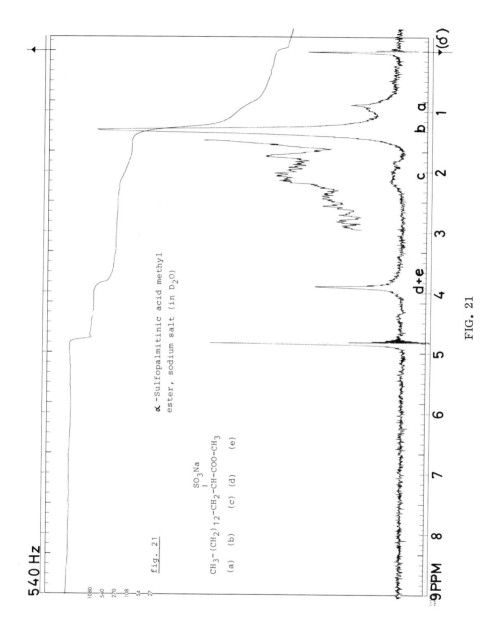

540 Hz

1080
540
270
108
54
27

fig. 21

α-Sulfopalmitinic acid methyl
ester, sodium salt (in D₂O)

$$CH_3-(CH_2)_{12}-CH_2-\overset{\overset{\displaystyle SO_3Na}{|}}{CH}-COO-CH_3$$

(a)   (b)   (c) (d)   (e)

FIG. 21

9 PPM   8   7   6   5   4   3   2   1   (δ)

d+e   c   b a

181

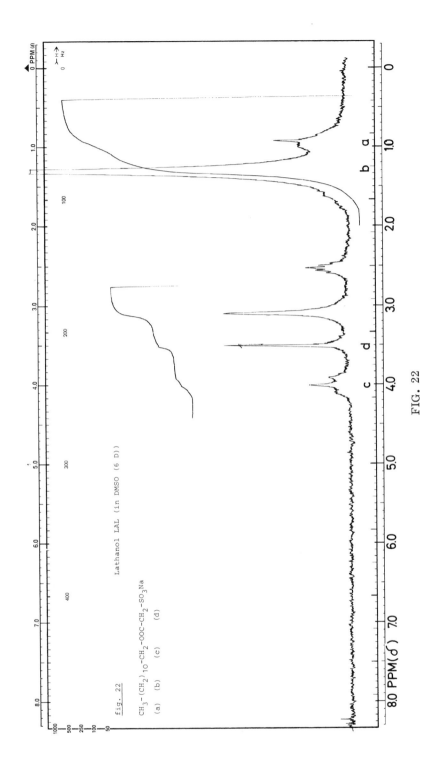

Lathanol LAL (in DMSO (6 D))

fig. 22

$CH_3-(CH_2)_{10}-CH_2-OOC-CH_2-SO_3Na$
(a) (b)   (c)   (d)

FIG. 22

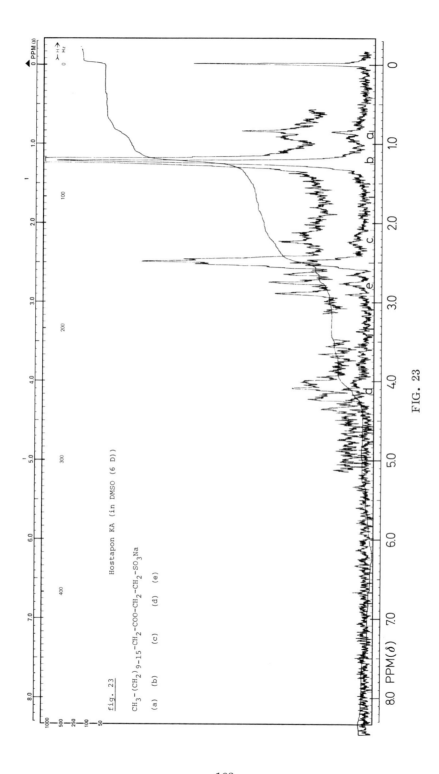

fig. 23   Hostapon KA (in DMSO (6 D))

$CH_3-(CH_2)_{9-15}-CH_2-COO-CH_2-CH_2-SO_3Na$

(a)   (b)   (c)   (d)   (e)

FIG. 23

183

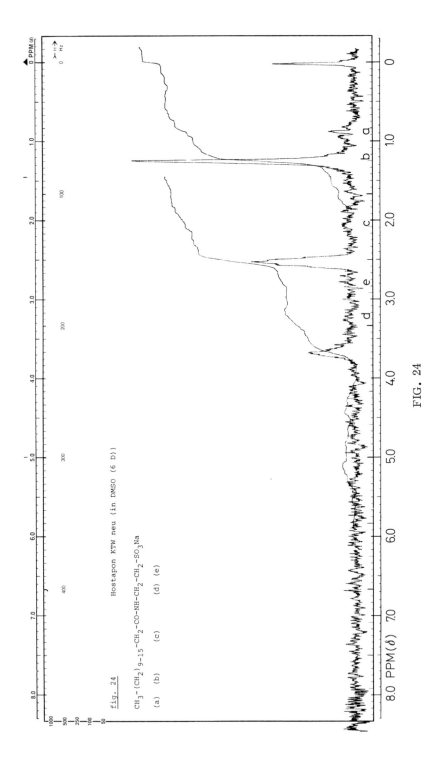

Hostapon KTW neu (in DMSO (6 D))

fig. 24

$CH_3-(CH_2)_{9-15}-CH_2-CO-NH-CH_2-CH_2-SO_3Na$

(a)   (b)   (c)   (d) (e)

FIG. 24

184

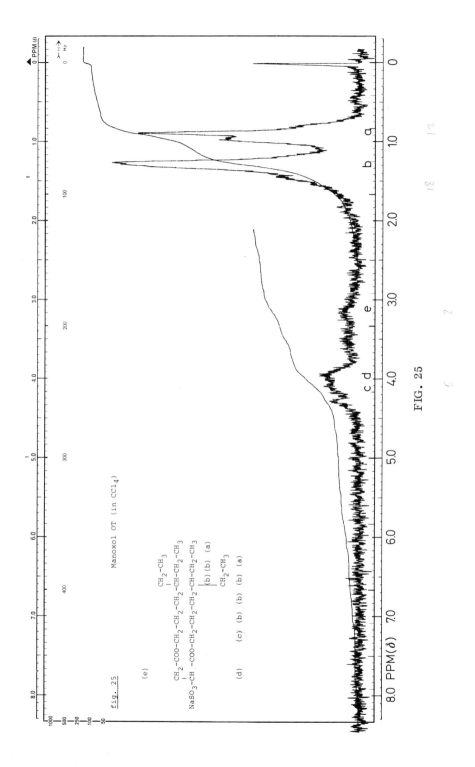

FIG. 25

Manoxol OT (in CC1₄)

(e)

$$CH_2-CH_3$$
$$|$$
$$CH_2-COO-CH_2-CH_2-CH-CH-CH_2-CH_3$$
$$|$$
$$NaSO_3-CH -COO-CH_2-CH_2-CH_2-CH-CH_2-CH_3$$
$$\text{(b) (b) (a)}$$
$$|$$
$$CH_2-CH_3$$
$$\text{(d)} \quad \text{(c) (b) (b) (a)}$$

fig. 25

185

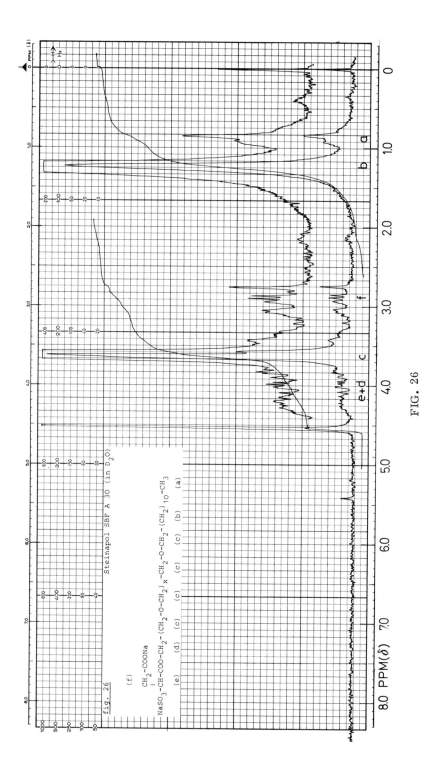

FIG. 26

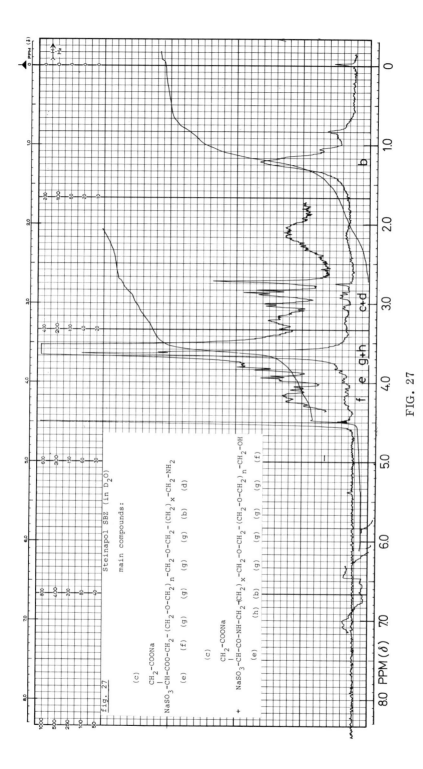

Steinapol SBZ (in D₂O)

main compounds:

(c)
|
CH₂-COONa
|
NaSO₃-CH-COO-CH₂-(CH₂-O-CH₂)ₙ-CH₂-O-CH₂-(CH₂)ₓ-CH₂-NH₂
(e)   (f) (g)   (g)   (g)   (g)   (b)   (d)

(c)
|
CH₂-COONa
|
NaSO₃-CH-CO-NH-NH-(CH₂)ₓ-CH₂-O-CH₂-(CH₂-O-CH₂)ₙ-CH₂-OH
(e)   (h) (b)   (g)   (g)   (g)   (f)
+

fig. 27

FIG. 27

187

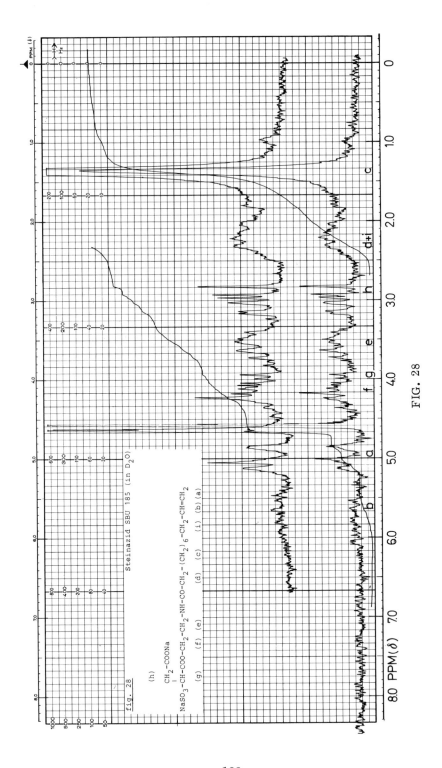

Steinazid SBU 185 (in D$_2$O)

$$CH_2-COONa$$
$$|$$
$$NaSO_3-CH-COO-CH_2-CH_2-NH-CO-CH_2-(CH_2)_6-CH_2-CH=CH_2$$
(g) (f) (e)   (d) (c)   (i) (b) (a)

(h)

fig. 28

FIG. 28

188

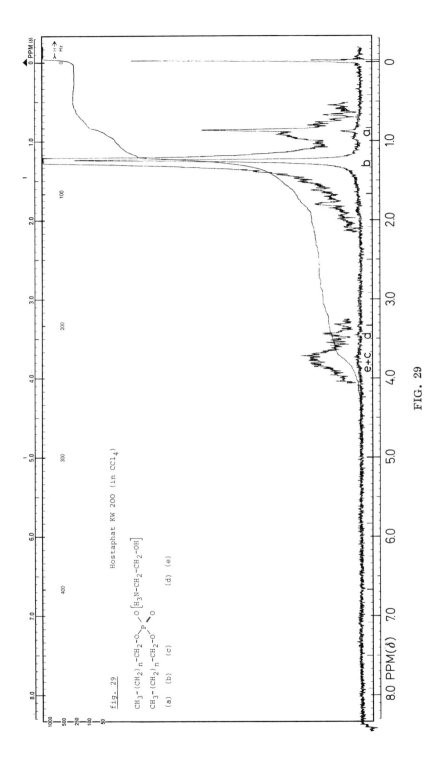

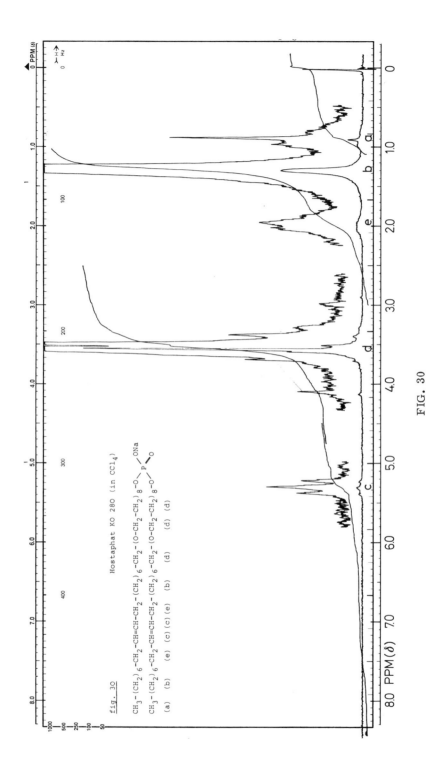

FIG. 30

fig. 30

Hostaphat KO 280 (in CCl₄)

$CH_3-(CH_2)_6-CH_2-CH=CH-CH_2-(CH_2)_6-CH_2-(O-CH_2-CH_2)_8-O$
$CH_3-(CH_2)_6-CH_2-CH=CH-CH_2-(CH_2)_6-CH_2-(O-CH_2-CH_2)_8-O$ P
$\begin{array}{c} ONa \\ O \end{array}$

(a) (b)   (e)(c)(e)   (b) (d) (d)

REFERENCES

1. R. R. Ernst and W. A. Anderson, Application of Fourier transform spectroscopy to magnetic resonance. Rev. Sci. Instr. 37, 93-102 (1966).
2. C. C. Hinckley, Paramagnetic shifts in solutions of cholesterol and the dipyridine adduct of trisdipivalomethanatoeuropium(III). A shift reagent. J. Am. Chem. Soc. 91, 5160-5162 (1969).
3. J. K. M. Sanders and D. H. Williams, A shift reagent for use in nuclear magnetic resonance spectroscopy. A first-order spectrum of n-hexanol. Chem. Commun., 422-423 (1970).
4. J. K. M. Sanders and D. H. Williams, Tris(dipivalomethanato)europium. A paramagnetic shift reagent for use in nuclear magnetic resonance spectroscopy. J. Am. Chem. Soc. 93, 641-645 (1971).
5. G. E. Stolzenberg, R. G. Zaylskie, and P. A. Olson, Nuclear magnetic resonance identification of o,p-isomers in an ethoxylated alkylphenol nonionic surfactant as tris(2,2,6,6-tetramethylheptane-3,5-dione)europium(III) complexes. Anal. Chem. 43, 908-912 (1971).
6. H. Walz and H. Kirschnek, Nuclear resonance spectroscopy as a valuable complement to infrared and ultraviolet analysis of surface-active compounds. Proc. 3rd Intern. Congr. Surface Active Substances, Köln, 1960, Vol. 3, pp. 92-98.
7. A. R. Greff, Jr. and P. W. Flanagan, The characterization of nonionic surfactants by NMR. J. Am. Oil Chemists' Soc. 40, 118-120 (1963).
8. M. M. Crutchfield, R. R. Irani, and J. T. Yoder, Quantitative applications of high-resolution proton magnetic resonance measurements in the characterization of detergent chemicals. J. Am. Oil Chemists' Soc. 41, 129-132 (1964).
9. T. F. Page, Jr. and W. E. Bresler, End-group analysis and number-average molecular weight determination of some polyalkylene glycols and glycol polyesters using nuclear magnetic resonance spectroscopy. Anal. Chem. 36, 1981-1985 (1964).
10. C. K. Cross and A. C. Mackay, Analysis of alkyl ethoxylates by NMR. J. Am. Oil Chemists' Soc. 50, 249-250 (1973).
11. H. König, Untersuchungen an Tensiden mit Hilfe der Kernresonanzspektroskopie. Z. Anal. Chem. 25, 225-262 (1970).
12. H. König, Neuere Methoden zur Analyse von Tensiden. Springer, Berlin, Heidelberg, New York, 1971, pp. 75-98.
13. T. Nagai, I. Tamai, S. Hashimoto, I. Yamane, and A. Mori, Olefin sulfonates. VI. Determination of 1-alkene sulfonate in alkene sulfonates by nuclear magnetic resonance spectroscopy. Kogyo Kagaku Zasshi 74, 32-35 (1971).

14. D. F. Kuemmel and S. J. Liggett, Level and position of unsaturation in alpha olefin sulfonates. J. Am. Oil Chemists Soc. <u>49</u>, 656-659 (1972).

15. H. König, Strukturaufklärung von Tensiden mit Hilfe der Kernresonanzspektroskopie. Tenside <u>8</u>, 63-65 (1971).

Chapter 5

THE COLORIMETRIC ESTIMATION OF ANIONIC SURFACTANTS

J. Waters   C. G. Taylor

Bioconsequences Section   Department of Chemistry
Unilever Research   Liverpool Polytechnic
Port Sunlight Laboratory   Liverpool, United Kingdom
Port Sunlight, Wirral
Merseyside, United Kingdom

I. INTRODUCTION

Colorimetric analytical procedures are routinely used to determine low levels of anionic surfactant in aqueous samples. All the colorimetric and many titrimetric procedures employed for determining anionic surfactants have a common analytical basis, that is, the formation of solvent-extractable compounds between the anionic surfactant and an intensely-colored cationic species. In some instances, colorimetric and titrimetric procedures have been developed with the same cationic dye reagent, e.g., methylene blue [1, 2].

Because of their high sensitivity the colorimetric methods are normally used to determine low levels of surfactant (10^{-5} to 10^{-8} M) in aqueous samples such as waste and surface waters, whereas the less sensitive titrimetric procedures are well suited to the determination of surfactant (10^{-2} to 10^{-4} M) in detergent pastes, powders, and finished products (see Chapter 6). Besides their advantage of high sensitivity the colorimetric procedures have been found to be quicker and simpler to operate and more free from serious interferences than any other techniques so far investigated for determining low levels of anionic surfactant.

The most important applications of colorimetric methods are in environmental analysis, where they are routinely used to monitor levels of anionic surface-active material in effluents, particularly sewage, and in river and potable waters. They are also used to follow the biodegradation of anionic surfactants in set biological testing procedures. In this chapter, therefore, particular emphasis is laid on those colorimetric procedures that can be used for both pure- and waste-water samples.

The most commonly used cationic reagent for colorimetric determinations is the organic dye methylene blue [1]. Other cationic dye systems which have had more limited application include methyl green [3], azure A [4], toluidine blue-O [5], rosaniline [6], and a range of iron(II) chelates of heteroaromatic polyamines [7, 8]. These cationic reagents are usually used in the form of simple salts such as the chloride or sulfate, which are not appreciably extracted from water by organic solvents. In the presence of anionic surfactant the cationic reagent can form a stable, stoichiometric ion-association compound (of much lower water solubility) by the neutralization of its positive charge with that on the surfactant polar group, i.e., sulfonate or sulfate. The resulting compound is readily extractable into organic solvents such as chloroform provided that the alkyl chain of the surfactant is sufficiently hydrophobic. The intensity of the color of the cationic-anionic surfactant compound in the extractant gives a direct measure of the surfactant present. The sensitivities of the colorimetric procedures are such that microgram amounts of anionic surfactant can be readily detected.

The cation-anion reaction is applicable to all the major anionic surfactant classes including alkylbenzenesulfonates (ABS), alkyl sulfates, alkanesulfonates, and alkenesulfonates. However, surfactants containing two or more polar groups, e.g., alkanedisulfonates, and the shorter-chain surfactants do not form readily extractable compounds [9]. In both cases the surfactant molecules are probably too hydrophilic.

When applied to mixtures of surfactants (different homologues and classes) colorimetric procedures give an estimate of the total surfactant present. Therefore, an appropriate reference surfactant has to be used in order to express results on a weight basis. Such standards include sodium dioctyl sulfosuccinate, Manoxol OT [10, 11], sodium lauryl sulfate, SLS [12, 13], sodium-1-phenyldodecane-p-sulfonate [14], and sodium tetra-propylenebenzenesulfonate, TBS [15]. Linear alkylbenzenesulfonate (LAS) and TBS have been commonly used as standards for waste-water analysis.

It is relatively easy to differentiate between alkyl sulfates and sulfonates in the same sample. Degens et al. [16] removed alkyl sulfates from samples, including sewage liquors, by an acid hydrolysis which left the sulfonate intact. A colorimetric estimation of the total surfactant present before and after hydrolysis enables both the sulfate and the sulfonate contents to be determined.

Cation-anion reactions cannot be specific for synthetic anionic surfactants, since any molecules containing a single strong anionic grouping and a hydrophobic part should also be capable of forming extractable compounds with cationic reagents. All colorimetric procedures are, therefore, susceptible to positive interferences, particularly from components present in natural waters. Negative interferences can also occur as a result of the direct competition from other cationic materials, such as proteins and quaternary ammonium compounds, with the cationic reagent for the anionic surfactant. Since all the procedures are based on the same type of reaction they can be expected to show similar interferences, the degree of interference being dependent upon the cationic reagent and the procedure used. The interference of some species can be partially or completely removed by the correct choice of reaction conditions or by separation. However, not all interferences can be eliminated, and the results obtained for unknown samples cannot be taken as representing anionic surfactant alone unless they have been checked against a reference method [17].

Although a number of other techniques such as polarography [18], fluorescence, atomic absorption [19], turbimetry and radiometry [20] can be used to determine microgram amounts of anionic surfactant, none of them have yet gained wide acceptance. Generally these other methods do not have the advantages associated with colorimetric procedures, i.e., simplicity, speed, high sensitivity, relative freedom from interferences, and wide applicability.

II. ORGANIC CATIONIC DYE SYSTEMS

Numerous organic cationic dyes and extractants have been investigated for determining anionic surfactant, but only a few have been found suitable [3, 21, 22]. Of the proven systems, methylene blue-chloroform is by far the most commonly used, whereas methyl green (benzene and chloroform), azure A (chloroform), toluidine blue-O (chloroform), and rosaniline (chloroform) have found more limited applications.

A. Methylene Blue

Jones [1] first proposed the use of methylene blue as a colorimetric reagent
for anionic surfactant. Since then numerous improved versions of his procedure have been developed. The continued popularity of methylene blue
must be ascribed to its high sensitivity and wide applicability, together with
a desire of analytical chemists to agree on a single reaction so that results
can be compared between laboratories.

1. Scope of Methylene Blue Reaction

Methylene blue [1] forms blue 1:1 compounds with all the major surfactant
classes. These compounds are extractable into chloroform provided the
alkyl chain of the surfactant is long enough. Alkylbenzenesulfonates with
less than five to six carbon atoms in the alkyl chain do not form chloroform-
extractable compounds with methylene blue. Sulfocarboxylated derivatives
of anionic surfactants, formed as a result of microbial breakdown, are
reactive to methylene blue when extracted under acidic conditions but are
less readily extracted than the corresponding noncarboxylated surfactants
of the same chain length [23]. Soaps are not responsive to methylene blue
under normal reaction conditions because the weak carboxylate anion is
incapable of forming stable compounds.

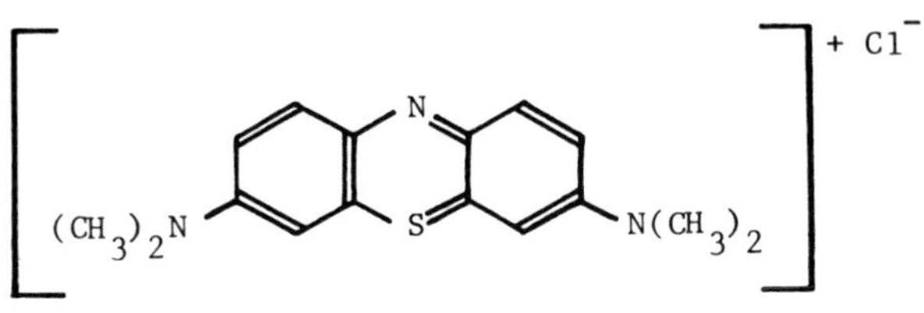

[1]

2. Interferences

Many natural organic materials, including sulfates, sulfonates, and some
simple inorganic anions such as cyanate, nitrate, thiocyanate and sulfide,
can form extractable compounds with methylene blue which interfere positively with the determination of anionic surfactant [24, 25]. Some organic
compounds, especially proteins and quaternary ammonium compounds, can
compete with the methylene blue for the surfactant and cause low results
[21]. In practice, positive errors are much more common than negative
ones, particularly for natural waters where the degree of interference depends largely upon the procedure used. The direct methylene blue results
for unknown samples are correctly referred to as methylene blue–active
substances (MBAS) and not as total anionic surfactant. Further investigation
is necessary to distinguish between the two.

3. Procedures

   a. Direct Procedures. Jones [1] completely extracted methylene
blue-anionic surfactant compounds into chloroform (four extractions) from
an acidified solution containing an excess of methylene blue. Each chloroform extract was washed once with distilled water to remove traces of
methylene blue salts. The absorbance of the combined chloroform extracts
was measured at 652 nm and the surfactant in the sample (100–400 μg; 20
ml sample) was determined by reference to a standard calibration.

   The procedure was capable of determining as little as 1 ppm of anionic surfactant (standard deviation, S.D. ±1% at the 10 ppm level) in distilled water, but was found to be subject to serious interference when
applied to the determination of surfactant in surface and waste waters.
Evans [26] showed that nitrate, thiocyanate, and certain components of
urine (sulfate esters) in waste waters interfered appreciably with Jones'
procedure by forming methylene blue–extractable compounds. A 1.8%
sodium chloride solution at pH 1.8 was found to give a response equivalent
to 10 ppm of alkyl sulfate, as did 1000 ppm nitrate or 40 ppm thiocyanate.
Evans also found that the extraction of anionic surfactant was unaltered as
the pH of the aqueous solution was decreased, whereas the extraction of
inorganic anions and compounds from urine decreased. By performing two
separate methylene blue determinations (five chloroform extractions) at pH
3.25 and 0.7 and linearly extrapolating the results to pH −2, Evans was able
to determine the surfactant content of samples containing these interferences. Although this artifice yielded satisfactory results for waste-water
samples (error ±0.5 ppm), a single determination took a long time.

   Degens et al. [16] showed that these same interferences were either
minimized or eliminated by adjusting the volumes of the aqueous and chloroform phases and by reducing the number of extractions to three. Despite
these improvements the Degens procedure was still susceptible to the positive interference of inorganic anions and of many organic components of
natural waters. Both the above procedures gave low recoveries of added
anionic surfactant in the presence of proteinaceous material such as sewage
solids. In the Degens method recoveries of ABS added to sewage varied
from 80 to 90%.

   In an attempt to improve the methylene blue procedures based on extractions from acid solution, Longwell and Maniece [11] proposed a double-
extraction method in which the methylene blue–anionic compound was extracted from alkaline solution (pH 10, phosphate buffer) and the resulting
chloroform extract washed with an acid solution of methylene blue to remove
the methylene blue degradation products (extracted under alkaline conditions) and the interferences by salts. Three chloroform extractions were
used to ensure the complete extraction of all anion–active material in the
sample. The alkaline extraction step, which was carried out above the
isoelectric point for protein, largely eliminated their negative interference.
It also removed positive interference of certain components of waste waters
such as sulfocarboxylates (breakdown products of the original surfactants).

The acid wash completely removed the positive interference of inorganic anions (except sulfide). The Longwell and Maniece procedure enables very low levels of anionic surface active material to be determined (as little as 0.01 ppm in a 100-ml sample of effluents, and river and sea water) because of the removal of the positive interferences of inorganic anions. The procedure results also in higher recoveries of surfactant from sewage samples. In their procedure, Longwell and Maniece recommended that sodium dioctyl sulfosuccinate, a readily obtainable, pure, stable surfactant, should be used as the reference material for all determinations.

The Longwell and Maniece procedure is probably the most used of all the colorimetric methods for determining low levels of surfactant in pure and waste-water samples and has been adopted by the U.K. Standing Technical Committee on Synthetic Detergents as the official UK procedure (STCSD, 1956). An outline of the procedure is given in Appendix 1.

Abbott [27] proposed an improved version of the Longwell and Maniece method which is particularly applicable to river and potable waters containing trace amounts of anion-active material. He showed that the pre-extraction of reagents (alkaline conditions) results in smaller reagent blanks and an increased sensitivity of about 15% compared with the Longwell and Maniece procedure. The increased sensitivity is a result of the removal of chloroform-soluble degradation products present in commercial grades of methylene blue (methylene azures A and B and dimethylthionoline), which compete with methylene blue for anion-active material during the alkaline extraction and the acid wash, but have lower molar absorptivities. In the Abbott procedure a borate buffer is used for the alkaline extraction step, redistilled analytical-grade chloroform and deionized water for reagent preparation and washing. This ensures high sensitivity and reproducibility for low-level determinations. The small reagent blanks permit the use of 4-cm optical cells to determine less than 0.01 ppm (1 μg in 100 ml) of surfactant in surface and potable waters.

The Abbott procedure (described in Appendix 2) is probably the most accurate and sensitive of all the direct methylene blue procedures, but the least convenient. Because the results obtained by this method are similar to those obtained by the Longwell and Maniece method, the latter is usually preferred because of its greater sample handling rate.

Although double-extraction procedures eliminate most of the positive and negative interferences, they do not remove those of sulfonates and sulfates found in natural waters [24, 25]. Therefore, when applied to natural waters the procedures give an estimate of the total anion-active material rather than the true synthetic syrfactant concentration.

b. Indirect Procedures. Fairing and Short [24] and Webster and Halliday [25] developed indirect methylene blue procedures so that ABS, the major commercial anionic surfactant, can be accurately determined in surface and waste waters. In both procedures the ABS is selectively extracted as its 1-methyl-n-heptylamine salt in order to isolate it from interfering materials. Fairing and Short first removed the ABS from substances such as proteins, which cause negative interferences, by an amine-chloroform extraction from neutral solution (pH 7.5). Interfering organic sulfates and other hydrolyzable components were then removed by acid hydrolysis. Finally, the ABS was isolated from substances causing positive interferences such as natural aliphatic sulfates and sulfonates by an amine-hexane extraction from acid solution (pH 4.8). In the Webster and Halliday procedure both amine extractions were carried out at a pH of 7.5 after an initial acid hydrolysis stage. In each case the isolated ABS, freed from traces of amine, was determined by a methylene blue procedure.

The Webster and Halliday procedure, applied to surface and waste waters, gave results which were 35-40% lower than those obtained by the Longwell and Maniece procedure (see Table 1). Results from the former procedure, however, showed close agreement with the infrared method specific for the determination of ABS [17].

The extra surfactant material found by the direct methylene blue procedure was a direct measure of the positive interference of natural anion-active material. Urine, aqueous extracts of vegetation, and uncontaminated water samples, which were shown to contain little or no ABS by the Webster and Halliday procedure, gave significant apparent surfactant levels by the Longwell and Maniece procedure (see Table 2).

The indirect methylene blue procedures are very accurate for determining ABS in surface and waste waters, but are lengthy and difficult to operate; a single determination can take up to 3 hr. As a consequence they are never used routinely but only to check unusual results obtained on samples analyzed by one of the direct procedures. For most environmental monitoring purposes the latter are adequate and are simple to operate.

The salient features of modifications to the methylene blue method are given in Table 3. All these modifications can be used to accurately determine ppm levels of anionic surfactant in distilled water, but only certain of them are effective for surface and waste waters.

4. Precision and Accuracy

The accuracy and precision of all the colorimetric procedures are dependent upon the surfactant to be analyzed, its concentration, and the composition of the sample [28, 29]. The most accurate and precise results are always obtained for determinations in the absence of interfering materials as, for example, in distilled water. The poorest results are obtained with sewage samples where representative sampling is an additional problem. For environmental samples, where biodegradation of the anionic surfactant can occur, immediate analysis or efficient sample preservation is essential to ensure the consistency of the results [30].

Gameson [31] has analyzed the results obtained by an interlaboratory test (16 participating laboratories) organized to examine the precision and accuracy of the Longwell and Maniece methylene blue procedure when

TABLE 1

Comparison of Methods on Surface and Waste Waters

Sample	Surfactant found, expressed as ppm of Manoxol OT		
	Direct methylene blue method[a]	Indirect methylene blue method[b]	Infrared method[c]
River water			
1	1.05	0.67	0.68
2	1.84	1.35	—
3	0.23	0.08	—
4	0.13	0.06	—
5	0.04	nil	—
Tap water	0.05	nil	—
Final sewage effluent			
1	2.9	2.6	2.5
2	2.5	1.7	1.7
3	2.4	1.9	1.8
Settled sewage			
1	4.6	3.6	—
2	14.0	9.6	—
3	12.0	9.9	—
4	14.7	9.7	—

[a] Longwell and Maniece [11].
[b] Webster and Halliday [25].
[c] Sallee et al. [17].

applied to natural water samples containing Manoxol OT, and branched and linear ABS. He observed a wide variability in results and concluded that this was due to systematic differences in operating the method. Although the reproducibility of results was good within a single laboratory, wide variations were found between laboratories. The greatest variability was observed for sewage samples, a settled sewage sample containing 9 ppm

TABLE 2

Interferences Found for the Direct and
Indirect Methylene Blue Methods

| Sample | Apparent surfactant found, expressed as ppm Manoxol OT | |
	Direct methylene[a] blue method	Indirect methylene[b] blue method
Urine	2.46	0.02
Aqueous hay extract	0.21	0.01
Aqueous coffee extract	0.68	0.04
Aqueous tea extract	0.16	nil
Peaty water A	0.13	nil
Peaty water B	0.11	nil
Peaty water C	0.18	nil
Extract of river bed vegetation	0.77	0.08

[a] Longwell and Maniece [11].
[b] Webster and Halliday [25].

of surfactant was in error by as much as ± 3 ppm, and an effluent containing 2 ppm was in error by ± 1 ppm. Generally, relative standard deviations were about ± 10–20%.

An interlaboratory exercise [32] organized by the Analytical Reference Service of the U.S. Public Health Service (A.R.S.) examined the precision and accuracy of three methylene blue procedures, the British Abbott method [27], the American (APHA) Standard method [33], and the American SEC method [34]. Distilled water and sewage containing 0.4 and 7.45 mg/liter of ABS, respectively, were analyzed. Nonnormal distributions of results were obtained for all three methods indicating doubtful interlaboratory reliability. Of the procedures investigated, the British method was considered the most accurate and precise (see Table 4).

Interlaboratory values for the precision of the American Standard Method are regularly reported by APHA [35, 36]. The latest are given in Table 5.

The above and similar studies [29, 37] indicate that the interlaboratory agreement for the commonly used procedures is not particuarly good, but

TABLE 3

Salient Features of the Colorimetric Methylene Blue Modifications

Method	Extraction conditions	Number of extractions	Methylene blue used per determination, mg	Approximate analysis range, μg	Maximum sample size, ml	Applicability
Jones [1]	acid	4	1.0	100–400	20	Distilled water
Degens et al. [16]	acid	3	1.75	10–250	10	Surface waters
American Standard method [36]	acid	4	0.8	10–200	400	Surface waters
Longwell and Maniece [11]	1. alkaline 2. acid	3	1.75 1.75	20–150	100	Surface and waste waters
Abbott [27]	preextraction of reagents			0–200	100	Surface and waste waters
	1. alkaline 2. acid	3	1.25 1.25			

TABLE 4

Standard Deviations for Three Methylene Blue Methods

Method	Sample	Surfactant, concn. in ppm	Number of determinations	Mean	Standard deviation, %
British [27]	Distilled water	0.4	42	0.41	±11.2
American Standard method [33]		0.4	57	0.42	±14.8
American SEC [34]		0.4	56	0.42	±14.8
British [27]	Sewage	7.45	43	7.42	±13.6
American Standard method [33]		7.45	57	7.21	±14.4
American SEC [34]		7.45	56	6.71	±15.7

TABLE 5

Precisions for the American Standard Method

Surfactant	Sample	ppm	Number of determinations	Standard deviation, %	Standard deviation for individual analyst, %
LAS	Distilled water	0.27	110	±14.8	—
	Tap water	0.48	110	± 9.9	—
	River water	2.94	110	± 9.1	—
ABS	Distilled water	0.40	57	±14.8	± 7.5
	Surface water	2.60	53	± 8.8	± 5.8
	Sewage	8.18	57	±12.7	± 3.1

that more reproducible results can be obtained by individual analysts (see Table 5). For most environmental monitoring purposes the precision and accuracy of the procedures are probably satisfactory.

5. Automation

The methylene blue reaction has been successfully adapted for the automatic colorimetric determination of anionic surfactants in aqueous samples using, for example, the Technicon Auto-Analyzer. Automation has resulted in higher analysis rates and increased precisions for surfactant determinations. Since the mode of operation of the manual methods is known to effect their variability [31], an increase in precision may be expected from the standardization by automation.

In automatic systems, the sample, methylene blue reagent, and chloroform are introduced separately into a continuous flow system by means of a proportioning pump. The methylene blue-anionic compounds are extracted into chloroform after passage through a series of mixing coils. The extract is either phase-separated from the excess aqueous reagent for color measurement, or is resampled for a further extraction step before measurement.

The simplest adaptations are based on the extraction of the methylene blue-anionic compounds from acidified methylene blue solutions and are particularly successful in determining anionic surfactant in distilled water, having high precision and little or no bias. Typically, 30 determinations per hour in the concentration range of 0.5 to 30 mg/liter can be performed [38, 39].

For surface and waste water determinations, adaptations of the Longwell and Maniece [11] double-extraction procedure (incorporating alkaline and acid methylene blue extractions) are normally employed in order to minimize the interferences associated with these samples [40, 41]. Södergren [40] showed statistically that there was more than a 90% probability of obtaining the same value for homogeneous samples containing 0.02-2.7 ppm of Manoxol OT by the automated and manual Longwell and Maniece procedure. An average analysis rate for these systems is about 20 determinations per hour, which is a considerable saving in time over the manual method. However, for sewage liquors containing solids the automatic procedures can sometimes give less accurate results than the manual because the small volumes of liquor required for a determination may not be representative of the bulk sample. In addition, low results may be produced because of incomplete desorption of surfactant from sewage solids, by the less vigorous automatic extraction systems.

The automation of the methylene blue reaction has resulted in great savings of time and labor. These systems are in regular use in water analysis laboratories for monitoring levels of anionic surfactant in environmental samples and for determining their biodegradability [42-47].

6. Field Tests and Comparative Color Procedures

The presence of anionic surfactant in normally uncontaminated surface and potable waters is an indication of pollution by waste water. Simple comparative methylene blue procedures, suitable for both field and laboratory use, have been developed for rapid spot testing of anionic surfactant in such samples. The procedures employ a single chloroform extraction of the surfactant from acidified methylene blue, which results in virtually complete removal of all surfactant. Only a portion of the chloroform phase need be taken and its color measured by visual comparison against a set of glass standards or specially prepared permanent copper-cobalt salt solutions equivalent to a range of known surfactant concentrations [48, 49]. Commercial field kits are available which enable as little as 0.05 mg/liter and as much as 2 mg/liter of anionic surfactant to be determined without dilution. The procedures have poorer precisions than the laboratory methods but no large biases [28]. They are susceptible to the same interferences as the laboratory methods (using acidic extraction conditions) but are generally free from the interference of most constituents found normally in surface and potable waters.

B. Methyl Green

Moore and Kolbeson [3] claimed that methyl green [2] is a superior reagent to methylene blue for determining anionic surfactant in natural waters because the interference from certain natural components is less. In their procedure, the surfactant (5-100 μg, 20 ml sample) is extracted once with benzene (40 ml) from an acidic solution (10 ml glycine-hydrochloric acid buffer, of pH 2.5) containing a large excess of methyl green (2 ml of 0.5% methyl green solution); the benzene extract is acid washed (15 ml of buffer) to remove any interfering water-soluble dye compounds. The absorbance of the methyl green-anionic compound is measured at 615 nm and the

[2]

surfactant content of the sample is obtained by reference to a prepared calibration (5-100 μg).

For distilled water determinations the method is accurate and precise (S.D. ±3% for 10 μg SLS). When compared with a methylene blue procedure, e.g., Jones [1], it was shown to be less susceptible to the interference of thiocyanate, nitrate, peptone, urea, and partially degraded surfactant [50]. However, the recovery of anionic surfactant is not quantitative in the presence of proteinaceous material (sewage solids) under the acidic extraction conditions employed.

The Moore and Kolbeson methyl green procedure is not widely used, because of the hazards of benzene and because of improvements in methylene blue methods, e.g., Longwell and Maniece [11].

More recently, Abbott [51] recommended the use of methyl green in a rapid field test for potable waters. A single chloroform extraction (5 ml) from acidic methyl green (5 ml of 2 N hydrochloric acid and 5 ml of 0.002% preextracted methyl green solution) made it possible to detect 1 μg of surfactant in a 200-ml sample in less than 5 min. The use of a reagent preextracted with chloroform to remove crystal violet impurity results in very low blanks. Similarly, the use of preextracted methyl green would enable benzene to be replaced by chloroform in the Moore and Kolbeson procedure.

C. Azure A

Steveninck and Riemersma [4] developed a rapid single-extraction procedure based on azure A [3]. The blue azure A-anionic compounds (wavelength of maximum absorption 630 nm) were found to be readily extracted into chloroform from acidic solutions. The procedure was very sensitive for long-chain alkyl sulfates in distilled water and at the same time was less susceptible than the Jones [1] methylene blue procedure to interference by thiocyanate and nitrate.

[3]

Tonkelaar and Bergshoeff [52] applied the procedure to the determination of trace levels of anionic surfactant (1-30 μg, 50 ml sample) in drinking and surface waters. They used the following procedure:

Dilute a suitable aliquot to 50 ml with water and add sulfuric acid
(5 ml, 0.05 M), 1 ml of azure A solution (0.04% in 0.0025 M sulfuric acid),
and chloroform (10 ml); shake for 2 min. Filter the chloroform layer
through a glass-wool plug into an optical cell (1 cm). Determine the absorbance at 623 nm against chloroform.

Extraction of the anionic-dye complex is confined to acidic solutions
because oxidation of the dye occurs in alkaline solution [27]. Tonkelaar and
Bergshoeff found that the low reagent blanks (0.005-0.010 in 1-cm cell) resulted in a highly sensitive method. Samples containing 0.04-0.16 mg of surfactant per liter were determined with a standard deviation of ±0.002 mg/
liter. The accuracy, however, was found to be dependent upon the presence
of interfering compounds and upon the surfactant used as reference standard.
The results obtained by the azure A and the Longwell and Maniece methylene
blue procedures were compared for several drinking and river waters.
Azure A gave consistently lower values, leading to the conclusion that it
might be less suceptible to naturally occurring organic compounds than
methylene blue. The azure A reaction appears to be a promising alternative, which is simple, rapid, and sensitive, for the determination of surfactant in potable waters.

Prins and Spaander [53] have claimed that azure A is applicable to
sewage determinations, but it is doubtful whether the negative interference
of proteins in sewage can be effectively eliminated under the acidic extraction conditions.

D. Toluidine Blue-O

A simple comparative field test, based on the cationic dye toluidine blue-O,
has been developed for determining anionic surface-active material in well-
water [5]. A single chloroform extraction (2.5 ml) of the anionic-dye compound from acidic solution (10 ml sample, 0.4 ml of 0.0001% toluidine
blue-O solution) is followed by a visual color comparison of the extract with
copper-cobalt color standards. The method detects as little as 0.05 ppm
of surfactant (range 0 to 20 μg). Components commonly encountered in
well waters do not seem to interfere. In practice, the method was found to
have a significant negative bias and a poorer precision than a comparable
methylene blue procedure [28]. The procedure had no obvious advantages
over existing field test methods.

E. Other Organic Dye Systems

A number of workers have investigated the possible use of rosaniline as a
colorimetric reagent for anionic surfactants [6, 54, 55]. The procedures
developed have low sensitivities, suffer from serious interference from
inorganic anions, and are of limited applicability.

III. ORGANOMETALLIC CATIONIC SYSTEMS

In the course of an investigation into methods of determining trace amounts
of iron in plutonium metal, Powell and Taylor [7] discovered that the cations
tris(bipyridine) iron(II) and tris(1,10-phenanthroline) iron(II), (ferroin),
could be extracted with chloroform after the addition of a small amount of
the anionic surfactant Teepol, any plutonium present remaining in the
aqueous phase. Mechanism studies showed that Teepol could be replaced
by many other anionic surfactants. Fryer [56] showed that the extraction
was due to the formation of stoichiometric ion-association compounds of the
type $[Fe(II)(chelate)_3]^{2+}[anionic]_2^-$. When the chelate is in formula excess
of the surfactant, the extraction of the chelate-anionic compounds is proportional to the anionic present. The spectral characteristics (Σ and
λ_{max}) of the chelate were found to undergo no change after extraction into
chloroform as the ion-association compounds.

Taylor and Fryer [13] developed a colorimetric method for determining anionic surfactant in sewage and sewage effluent based on the use of the
ferroin cation. Ferroin was used in preference to the bipyridine chelate
because it had the higher molar absorptivity (11,200 at 512 nm). The
ferroin-anionic compounds were completely removed from acetate buffered
solutions (pH 5), after the addition of an appropriate ferroin reagent, with
three chloroform extractions. For wholly liquid samples a dilute ferroin
reagent was used, whereas a concentrated reagent was necessary to give
quantitative recovery of surfactant from sewage. A capillary extraction
device was used to separate the chloroform extracts from the aqueous layer;
this device made it possible to operate on a semimicro scale (5 ml sample,
2 ml extractant) without the restrictions normally imposed by the use of
separating funnels. The extract absorbances were measured at 512 nm,
and the anionic content of samples obtained by reference to a prepared calibration (SLS was recommended as a reference surfactant). A more detailed
outline of the procedure is given in Appendix 3.

The ferroin method appears to have several advantages over methylene
blue procedures with regard to speed (10-15 min per determination), econo-
my (semimicro scale), purity of reagents, and reagent blanks. On the
other hand, ferroin is considerably less sensitive than methylene blue; the
estimated limit of detection for the method is 2.5 μg of SLS (0.5 ppm in a
5 ml sample). Nevertheless, it is well-suited to the concentration range of
surfactants normally found in sewage and sewage effluent.

The ferroin method and the Longwell and Maniece methylene blue procedure were concurrently applied to sewage samples, and a qualitative
agreement between them was obtained. However, the methylene blue results were usually higher than the ones obtained with ferroin. The precisions of the methods were comparable. Of all the inorganic ions examined
only thiocyanate seriously interferes. Its interference can be eliminated if
ferroin is replaced by bipyridine iron(II) chelate. Ferroin and its bipyridyl

analogue, like the organic dye reagents, are responsive to sulfonates and sulfates of natural origin that may be present in waters.

Taylor and Waters [20] developed a more sensitive radiometric modification of the original procedure by using ferroin labeled with [59Fe] to determine trace amounts of anionic surfactant in surface and potable waters. The method employs a single extraction since this is more rapid and also gives smaller blanks and a better limit of detection than multiple extraction. The [59Fe]ferroin-surfactant compounds are extracted from a solution buffered to pH 4 (0.5 ml of 2 M acetate/acetic acid, 20 ml sample) and containing an excess of ferroin (0.5 ml of 8.95×10^{-5} M ferroin, 0.5μCi, iron:1,10-phenanthroline ratio of 1:6) with water-equilibrated chloroform (5 ml). The extract is centrifuged to remove any entrained aqueous phase and transferred to a graduated counting vessel (approx. 4 ml) using the extraction device. The γ-activity of the extract is measured by scintillation counting. An internal standard technique is used for calibration.

The radiometric method is capable of determining 0.1 μg of surfactant (0.005 ppm in a 20-ml sample) with a precision better than $\pm 10\%$. Potential interferences of fluoride and iron, Fe(II) and Fe(III), are compensated for by the use of an internal standard; none of the other inorganic species studied interfered when present at the levels normally encountered in river and potable waters. The results of this method are in good agreement with the results of Abbott's methylene blue procedure, and it is somewhat more sensitive. It appears to be well-suited for the examination of natural waters for trace pollution by anion-active materials or for any other situation where there is a need to follow changes in small anionic surfactant concentrations.

Several other iron(II) chelates, chemically similar to ferroin, have been shown to form colored, stoichiometric, chloroform-extractable compounds (listed in Table 6) with anionic surfactants [8]. The most important feature of this series of organometallic cations is that in each surfactant series there is a member which forms a partially-extractable compound with a particular chelate, approx. 50%, termed the "break-through compound" (see Table 6). An increase in alkyl chain length of the surfactant above that of the break-through compound results in increased extractability, and vice versa. The break-through chain length is also dependent on the surfactant series and the structure of the chelate (number and type of substituents). The largest chelates form the most extractable compounds and the smallest the least extractable ones for a particular surfactant.

The conditional extractability of many of these compounds was measured by Taylor and Waters [57] and expressed as their extraction constants. Appreciable differences were found between the extraction constants of compounds of adjacent homologous surfactants with a particular chelate. Taylor and Waters [58] investigated the factors which influence the extraction and separation of these chelate-anionic homologous compounds and developed methods for determining (a) homologous surfactants of various

TABLE 6

Break-Through Compounds of Iron(II) Chelate-Surfactant Series

Ligand	Break-through alkyl chain length		
	Alkane-sulfonate	Alkyl sulfate	Alkylbenzene-sulfonate
2,2'-Bipyridine	C_{12}	C_{10}	C_8
2,2',2''-Tripyridine	C_{12}	C_{10}	C_8
1,10-Phenanthroline	C_{10}	C_8	C_6
5-Nitrophen-1,10-phenanthroline	C_{12}	C_8	C_6
5-Chloro-1,10-phenanthroline	C_8	C_6	C_4
5-Bromo-1,10-phenanthroline	C_8	C_6	C_4
5-Methyl-1,10-phenanthroline	C_8	C_6	C_4
(4,7- or 5,6-)Dimethyl-1,10-phenanthroline	below C_8	C_5	below C_4

chain lengths and (b) short-chain surfactants. Among the former, three surfactant series can be fractionated over a range of seven carbon atoms (i.e., C_3-C_{10} ABS, C_5-C_{12} alkyl sulfate, and C_7-C_{14} alkane sulfonate) by the use of four metal chelate cations and two extractants (chloroform and 15% chloroform in carbon tetrachloride). In the latter, a 4,7-dimethyl-1,10-phenanthroline chelate reagent is used to determine μg amounts of short-chain surfactant in solution down to C_3-ABS, C_5-alkyl sulfate and C_7-alkanesulfonate. The methods have been successfully applied to synthetic mixtures, river waters, and synthetic sewage liquors.

IV. CONCENTRATION AND SEPARATION
     TECHNIQUES

For most determinations the colorimetric methods are sensitive enough to estimate the anionic content of aqueous samples when only small volumes are available. However, in certain instances where the values to be determined are near the blank value for the colorimetric method, the surfactant in the sample must be concentrated before an accurate estimation can be made. Generally, concentration by simple evaporation is rejected because of its slowness and nonspecificity. Three techniques have been successfully used for concentrating surfactants with some separation from other materials, that is, activated carbon column chromatography, ion exchange, and solvent sublation.

Sallee et al. [17] quantitatively adsorbed anionic surfactant onto purified Nuchar-C 190 activated carbon (100 g) from large volumes of surface water (up to 20 liters). The concentrated surfactant was desorbed by boiling the dried activated carbon in an alkaline benzene-methanol mixture. Recoveries of greater than 95% of surfactant were obtained with this technique from samples containing as little as a few ppb. Ogden et al. [59] modified the procedure so that the adsorbed surfactant was removed by elution of the activated carbon with a methanol-chloroform-ammonia solvent mixture, thus speeding up the process and reducing the solvent requirements by one quarter. Although this technique is used as the first step in infrared reference procedures for determining ABS in surface and potable waters [17, 59], it is too time and solvent consuming to be used routinely.

Concentration by ion-exchange chromatography is more convenient since little resin preparation is needed and relatively small volumes of eluent are required to give quantitative recoveries of surfactant. Le Peintre and Romens [60] concentrated anionic surfactant by fixing it on Amberlite IRA 68 resin (hydroxide form). The surfactant was desorbed by eluting the resin with a mixture of acetone and 0.1 N sodium hydroxide, to give 95–100% recoveries. The accuracy of the procedure is about 10% for samples containing more than 2 μg/liter and 15–20% for those containing 1–2 μg/liter. Hughes [61] used Biorad AG1-X2 resin to concentrate anionic surfactant in effluents. Recoveries of 95% of anionic surfactant were obtained when the resin was eluted with methanolic hydrochloric acid (50 ml).

A most interesting technique is solvent sublation which is both rapid and simple to operate. Wickbold [62] showed that surfactant is quantitatively transported from acidified aqueous samples into an overlying layer of organic solvent by bubbling gas (nitrogen or air) through it. The phenomenon is due to adsorption of the surfactant on the gas bubbles and its subsequent transfer to the liquid-liquid interface where the isolated substances are dissolved. When ethyl acetate was used as the solvent and also to saturate the gas stream, quantitative recovery of ppb levels of surfactant were obtained after two separate sublation steps. Up to 5 liters of sample can be handled at one time with this technique. An outline of the technique is given in Appendix 4. Concentration of surfactant by sublation appears to remove certain materials which interfere positively in the Longwell and Maniece methylene blue procedure [62].

APPENDIX 1
The Methylene Blue Procedure of Longwell and Maniece [11]

This is the official U.K. procedure for the determination of anion-active material in sewage, sewage effluent, and river waters.

Reagents

Alkaline phosphate solution: Dissolve disodium hydrogen phosphate (10 g anhydrous) in distilled water. Adjust the pH to 10 by addition of sodium hydroxide and make up to 1 liter.

Neutral methylene blue solution: Dissolve reagent-grade methylene blue (0.35 g) in water and make up to 1 liter.

Acid methylene blue: Dissolve reagent-grade methylene blue (0.35 g) in water (500 ml), add sulfuric acid (18 M, 6.5 ml), and dilute to 1 liter.

Reference surfactant: Prepare 1000 ppm and 10 ppm solutions of Manoxol OT.

Procedure

Place the sample containing 20-150 μg of anionic surfactant in a separating funnel and make up to 100 ml with distilled water. (Generally take 10 ml of settled sewage, 50 ml of sewage effluent, or 100 ml of river or potable water for analysis.) Add alkaline phosphate solution (10 ml), neutral methylene blue solution (5 ml), and chloroform (15 ml). Shake gently (1 min). Allow layers to separate. Run the clear chloroform layer into a second separating funnel containing distilled water (110 ml) and acid methylene blue (5 ml). Rinse the first separating funnel with chloroform (2 ml) and run this into the second separating funnel. Shake the second separating funnel (1 min) and allow the layers to separate. Run the chloroform layer through a small funnel plugged with cotton wool moistened with chloroform into a graduated flask (50 ml). Repeat the complete extraction step twice more with portions of chloroform (10 ml) and make up the combined extracts to the mark.

Measure the absorbance of the extract (650 nm, 1-cm cell) against a chloroform reference. Subtract the reagent blank from the sample reading and determine the anionic content of the sample by reference to a prepared calibration (range 0-200 μg).

APPENDIX 2
The Methylene Blue Procedure of Abbott [27]

This procedure is particularly applicable to the determination of small amounts of anion-active material in river and potable water.

Reagents

Alkaline borate solution: Mix equal volumes of sodium tetraborate (0.05 M) and sodium hydroxide (0.10 N).

Sulfuric acid (approximately 0.5 M).

Methylene blue solution: Dissolve B.P. or "vital-stain" grade material (0.25 g) in water and dilute to 1 liter.

Chloroform: This must be free from ethanol and surfactant. Wash the
chloroform twice with dilute (1:100) methylene blue solution, distill over
burnt lime, and filter the distillate through anhydrous sodium sulfate.

Reference surfactant: Prepare 1000 and 10 ppm solutions of Manoxol OT.

Deionized water should be used for reagent preparation.

Procedure

Measure into a 250-ml separating funnel water (50 ml), alkaline borate
solution (10 ml), and methylene blue solution (5 ml). Add chloroform (10
ml), shake (30 sec), and allow to separate. Draw off the chloroform layer
as completely as possible and rinse the aqueous layer, without shaking,
with chloroform (2-3 ml). Repeat the extraction with chloroform (10 ml),
and rinse as before.

Measure water (100 ml), alkaline borate solution (10 ml), and methylene blue (5 ml) into a second separating funnel. Extract twice with 10-ml
portions of chloroform as described above. Add sulfuric acid (0.5 M, 3 ml)
to the extracted aqueous layer in this funnel and mix well.

Place the sample (preferably not exceeding 100 ml and containing 50-
100 μg of anion-active material) into the first funnel. (Use 10 ml of settled
sewage, 50 ml of sewage effluent or 100 ml of river or drinking water.)
Add chloroform (15 ml) and shake evenly (1 min). Allow the layers to
separate (2 min) and then draw as much as possible of the chloroform layer
into the second separating funnel containing the acidified methylene blue.

Shake the contents of the funnel, and allow to separate (2 min). Run
the chloroform layer through a small filter funnel plugged with cotton wool
(freshly washed with chloroform) into a graduated flask (50 ml); allow none
of the aqueous phase to enter the tap of the separating funnel. Twice repeat
the operations above, from "Add chloroform (15 ml) ...," combine the
extracts in the flask, and dilute to the mark with chloroform.

Perform a blank determination in the manner indicated, and then
measure the absorbances of the reagent blank and sample chloroform solutions (650 nm, 1-cm cells) (4-cm cells for absorbances less than 0.1)
against a chloroform reference. Rinse the cell three times with the chloroform solution before reading the absorbance. Subtract the reagent blank
from the sample reading and determine the anionic content of the sample
by reference to a prepared calibration (0-200 μg).

APPENDIX 3
The Ferroin Method of Taylor and Fryer [13]

This is a semimicro procedure for determining anionic surfactant in sewage and sewage effluents.

Reagents

A standard reference SLS solution, 17.9×10^{-4} M.

Ammonium iron(II) sulfate solution, 17.9×10^{-3} M (1000 ppm of iron): Dissolve analytical-reagent grade $(NH_4)_2SO_4 \cdot FeSO_4 \cdot 6H_2O$ (7.023 g) in water, add hydroxylamine hydrochloride (1 g) and dilute to 1 liter.

Dilute ferroin reagent, 8.95×10^{-4} M (50 ppm with respect to iron): To the iron(II) solution (5 ml) add 1,10-phenanthroline monohydrate (0.1 g, 5×10^{-4} M); dissolve in water (25 ml) and dilute to 100 ml. This reagent contains a 5-6 molar excess of 1,10-phenanthroline with respect to iron(II).

Concentrated ferroin reagent, 8.95×10^{-3} M (500 ppm with respect to iron): To the iron(II) solution (50 ml) add 1,10-phenanthroline mono-hydrate (0.443 g); dissolve the solid and dilute to 100 ml. This reagent is for use with sewage samples containing suspended solids, and has a phenan-throline-to-iron(II) ratio of 2.5.

Dilute and concentrated iron(II)-bipyridine reagents for samples con-taining thiocyanate: Prepare at similar concentrations and ratios to those described above.

Buffer solution, pH 5: Adjust sodium acetate (2 M) to pH 5 with acetic acid (2 M).

Chloroform AR.

Procedure

The sample (5 ml) should be neutral, pH 4 to 10, and should contain not more than 150 μg of surfactant. Place it in a conical centrifuge tube (10 ml), add buffer solution (0.5 ml) followed by the appropriate ferroin reagent (1 ml). Use the dilute reagent if the sample is wholly liquid and the concen-trated reagent if it contains suspended solids.

Treat the solution with two successive portions of chloroform (2 ml) using the extraction technique described below, and combine the extracts in a graduated flask (5 ml). Wash the solution with chloroform (2 ml) and use the washing to dilute the extract to the mark.

Measure the absorbance of the extract (512 nm, 1-cm stoppered cells) against a chloroform reference. Perform a reagent blank following the above procedure.

Obtain the anionic content of the sample by reference to a previously prepared calibration graph (0-125 μg).

Extraction technique: Stir the contents of the tube mechanically at high speed for 1 min (the speed should be sufficient to mix the two layers completely), then centrifuge (1 min), after which the layers should be com-pletely clear. Remove the lower solvent layer from the tube to the graduated

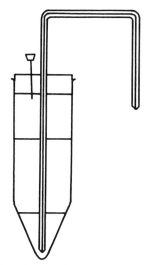

FIG. 1. Extraction device for removing chloroform layer.

flask by using the device shown in Fig. 1, as follows. Attach a teat to the short arm of the capillary and introduce the long arm through the aqueous into the solvent layer, maintaining a slight pressure on the teat to produce a slow stream of air bubbles. Secure the bung in the mouth of the tube, then remove the teat. Transfer the solvent to the flask by applying pressure with the syringe (5-10 ml) via the needle. When all but about 0.1 ml of solvent has been removed, disconnect the syringe, replace the teat, and withdraw the capillary, applying slight pressure to the teat as before.

There is little chance of water being transferred to the flask in this technique.

APPENDIX 4
The Solvent Sublation Technique of Wickbold [62]

This is a simple technique for concentrating anionic surfactant in surface and potable water.

Apparatus and Reagents

Ethyl acetate AR.
Gas-stripping apparatus (see Figure 2), 1 liter capacity.

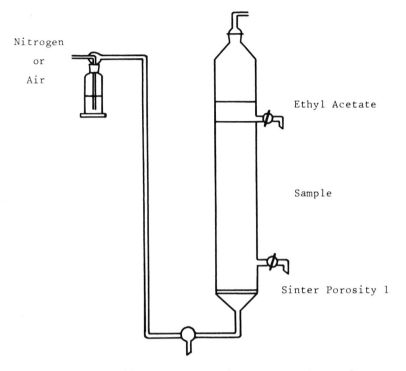

FIG. 2. Solvent sublation apparatus for concentrating surfactant.

Procedure

Acidify a sample (1 liter) with concentrated hydrochloric acid (10 ml) and
overlay it with ethyl acetate (100 ml). Fill the gas wash bottle with ethyl
acetate. Pass a current of gas (air or nitrogen) at a rate of 50-60 liters
per hour through the apparatus, at the same time ensuring that the solvent-
water interface is not unduly disturbed. Stop the gas supply after 5 min
and transfer the organic phase to a glass beaker. Cover the sample again
with ethyl acetate (100 ml), and pass gas through for another 5 min. Drain
the ethyl acetate layer off and evaporate the combined organic layers to
dryness. Determine the anionic surface-active material in the residue by
colorimetry.

REFERENCES

1. J. H. Jones, General colorimetric method for determination of small
   quantities of sulfonated or sulfated surface active compounds. J. Assoc.
   Offic. Agric. Chemists <u>28</u>, 398-409 (1945).

2. S. R. Epton, A new method for the rapid titrimetric analysis of sodium alkyl sulfates and related compounds. Trans. Faraday Soc. 44, 226-230 (1948).

3. W. A. Moore and R. A. Kolbeson, Determination of anionic detergents in surface waters and sewage with methyl green. Anal. Chem. 28, 161-164 (1956).

4. J. van Steveninck and J. C. Riemersma, Determination of long-chain alkyl sulfates as chloroform soluble azure A salts. Anal. Chem. 38, 1250-1251 (1966).

5. O. E. McGuire, F. Kent, L. L. Miller, and G. J. Papenmeier, Field test for analysis of anionic detergents in well waters. J. Amer. Water Works Assoc. 54, 665-670 (1962).

6. F. Karush and M. Sonenberg, Long-chain alkyl sulfates. Colorimetric determination of dilute solutions. Anal. Chem. 22, 175-177 (1950).

7. R. Powell and C. G. Taylor, The chloroform extraction of ferrous-dipyridyl and o-phenanthroline complexes with the aid of alkyl sulfates and sulfonates. Chem. Ind., 726 (1954).

8. J. Waters, Analysis of anionic surfactants using cationic chelates of iron (II). Ph.D. thesis, Liverpool Polytechnic (1971).

9. K. Ōba, A. Mori, and S. Tomiyama, Biochemical studies of n-α-olefine sulfonates. I. Biodegradability under aerobic condition. Yukagaku 17, 517-520 (1968).

10. H. L. Bolton and P. J. Cooper, Analytical problems in the Luton experiment. J. Proc. Inst. Sewage Purif., 43-47; discussion 49-56 (1961).

11. J. Longwell and W. D. Maniece, Determination of anionic detergents in sewage, sewage effluents and river waters. Analyst 80, 167-171 (1955).

12. V. W. Reid, G. F. Longman, and E. Heinerth, Determination of anionic detergents by two-phase titration. Tenside 4, 292-304 (1967).

13. C. G. Taylor and B. Fryer, The determination of anionic detergents with iron(II) chelates: Application to sewage and sewage effluents. Analyst 94, 1106-1116 (1969).

14. F. W. Gray, J. F. Gerecht, and I. J. Krems, Preparation of model long chain alkyl benzenes and study of their isomeric sulfonation products. J. Org. Chem. 20, 511-524 (1955).

15. V. W. Reid, G. F. Longman, and E. Heinerth, Determination of anionic detergents by two-phase titration. II. Tenside 5, 90-96 (1968).

16. R. N. Degens, Jr., H. C. Evans, J. D. Kommer, and P. A. Winsor, Determination of sulfate and sulfonate anion-active detergents in sewage. J. Appl. Chem. (London) 3, 54-61 (1953).

17. E. M. Sallee et al., Determination of trace amounts of alkyl benzene-sulfonates in water. Anal. Chem. 28, 1822-1826 (1956).

18. K. Linhart, Die polarographische Bestimmung von grenzflächenaktiven Stoffen in Wasser und Abwasser sowie die Bestimmung von deren Ab-Abaubarkeit. Tenside 9, 241-259 (1972).

19. J. Courtot-Coupez and A. Le Bihan, Determination of anionic detergents in sea water. Generalisation from the method for fresh water and for the determination of cationic detergents. Anal. Letters 2, 211-219 (Fr.) (1969).

20. C. G. Taylor and J. Waters, Radiometric determination of trace amounts of anionic surfactants in ground water and potable water. Analyst 97, 533-541 (1972).

21. G. P. Edwards and M. E. Ginn, Determination of synthetic detergents in sewage. Sewage Ind. Wastes 26, 945-953 (1954).

22. L. G. C. (Laboratory of the Government Chemist), Report of the Government Chemist 1962, HMSO, London, 1963.

23. R. D. Swisher, J. T. O'Rourke, and H. D. Tomlinson, Fish bioassays of LAS and intermediate biodegradation products. J. Am. Oil Chemists' Soc. 41, 746-752 (1964).

24. J. D. Fairing and F. R. Short, Spectrophotometric determination of ABS detergents in surface water and sewage. Anal. Chem. 28, 1827-1834 (1956).

25. H. L. Webster and J. Halliday, Determination of alkylbenzenesulfonates in river waters and sewage. Analyst 84, 552-559 (1959).

26. H. C. Evans, Determination of anionic synthetic detergents in sewage. J. Soc. Chem. Ind. (London), Suppl. Issue No. 2, 576-580 (1950).

27. D. C. Abbott, Colorimetric determination of anionic surfactants in water. Analyst 87, 286-293 (1962).

28. C. H. Wayman and A. T. Miesch, Accuracy and precision of laboratory and field methods for the determination of detergents in water. Water Resources Res. 1, 471-476 (1965).

29. H. L. Bolton, H. L. Webster, and J. Hilton, Collaborative work on the determination of ABS in sewage, sewage effluents, river waters and surface waters. J. Proc. Inst. Sewage Purif., 302-308 (1962).

30. R. D. Swisher, Surfactant Biodegradation, Surfactant Science Series Vol. 3, Marcel Dekker, Inc., New York, 1970.

31. A. L. H. Gameson and V. H. Lewin, The determination of anionic surface-active materials. J. Proc. Inst. Sewage Purif., 288-301 (1962).

32. E. E. Finnecy and N. J. Nicolson, Proc. Soc. Water Treat. Exam. 17, 8 (1968).

33. APHA (American Public Health Association, American Water Works Association, and Water Pollution Control Federation), Standard Methods for the Examination of Water and Waste Water, 11th ed., APHA, New York, 1960.

34. ARS, Analytical Reference Service of the U.S. Public Health Service; Robert A. Taft Sanitary Engineering Center. Analytical reference service; Water Surfactant No. 2. Cincinnati, The Center, 1964; vii + 105pp.

35. APHA, Standard Methods for the Examination of Water and Waste Water, 12th ed., APHA, New York, 1965.

36. APHA, Standard Methods for the Examination of Water and Waste Water, 13th ed., APHA, New York, 1971.
37. DAGS (Deutscher Ausschuss für Grenzflächenaktive Stoffe), Fachkommission "Abwasser." Critical observations on the utility of the Longwell-Maniece method. Gas-Wasserfach 52, 1426-1427 (1961).
38. A. L. de Jong, Determination of anionic surfactants with the Auto-analyser. Fette, Seifen, Anstrichmittel 71, 567-569 (1969).
39. M. Testa, Automatic analytical method for determining anionic detergents in polluted waters. Federation Europäischer Gewässerschutz, Informationsbl. 11, 33-36 (1964).
40. A. Södergren, An automatic method for the determination of anionic surface-active material in water. Analyst 91, 113-118 (1966).
41. A. W. Davies and K. Taylor, Application of the Auto-Analyser in a river authority laboratory. Technicon Fifth International Symposium, London, 1965.
42. R. Ellerker and B. Collinson, Use of an Auto-analyser for sewage work analyses. Water Pollution Control 71, 540-548 (1972).
43. E. L. Barnhart, Defining the degradability of synthetic detergents. Wastes Eng. 34, 646-648 (1963).
44. R. L. Huddleston and R. C. Allred, Evaluation of detergents by using activated sludge. J. Am. Oil Chemists' Soc. 41, 732-735 (1964).
45. C. E. Renn, W. A. Kline, and G. Orgel, Destruction of linear alkylate sulfonates in biological waste treatment by field test. J. Water Pollution Control Federation 36, 864-879 (1964).
46. E. A. Setzkorn, R. L. Huddleston, and R. C. Allred, An evaluation of the river die-away technique for studying detergent biodegradability. J. Am. Oil Chemists' Soc. 41, 826-830 (1964).
47. R. J. Kelly, M. S. Konecky, J. E. Shewmaker, and R. Bernheimer, Physical and biological removals of detergent actives in a full scale sewage plant. Surface activity and the living cell. Soc. Chem. Ind. (London), Monograph No. 19, 286-304 (1965).
48. Hellige, Inc., Technical Information Sheet 367-D, Garden City, New York, 1961, 2 pp.
49. E. Michelsen and E. Märki, Laboratory and field method for detection of anionic surfactants in surface, ground and waste waters. Mitt. Gebiete Lebensm. Hyg. 52, 557-571 (1961).
50. R. C. Allred, E. A. Setzhorn, and R. L. Huddleston, Detergent biodegradability as shown by various analytical techniques. J. Am. Oil Chemists' Soc. 41, 13-17 (1964).
51. D. C. Abbott, A rapid test for anionic detergents in drinking water. Analyst 88, 240-241 (1963).
52. W. A. M. den Tonkelaar and G. Bergshoeff, Use of azure A instead of methylene blue for determination of anionic detergents in drinking and surface waters. Water Res. 3, 31-38 (1969).
53. R. Prins and P. Spaander, The determination of anionic detergents by the azure-blue procedure. H$_2$O 1, 260-262 (1968).

54. G. R. Wallin, Colorimetric method for determining a surface-active agent. Anal. Chem. 22, 616-617 (1950).

55. R. W. G. Cropton and A. S. Joy, Determination of low concentrations of sodium dodecylbenzenesulfonate. Analyst 88, 516-521 (1963).

56. B. Fryer, Proc. Soc. Analyt Chem. 3, 44 (1966).

57. C. G. Taylor and J. Waters, Ion-association compounds of anionic surfactants with iron(II) chelates. Part I: Extraction constants. Anal. Chim. Acta 69, 363-371 (1974).

58. C. G. Taylor and J. Waters, Ion-association compounds of anionic surfactants with iron(II) chelates. Part II: Selective determination of surfactants. Anal. Chim. Acta 69, 373-387 (1974).

59. C. P. Ogden, H. L. Webster, and J. Halliday, Determination of biologically soft and hard ABS in detergents and sewage. Analyst 86, 22-29 (1961).

60. C. Le Peintre and C. Romens, Method of determinations of traces of anionic detergents in water. Compt. Rend. 261, 452-454 (1965) (Fr.); C.A. 63, 9660(a) (1965).

61. W. Hughes, S. Frost, and V. W. Reid, Analysis of alkylbenzene sulfonates present in sewage. International Congress on Surface Activity, Barcelona, 1968, Chimie, Physique et Applications Pratiques des Agents de Surface, Vol. 1. Ediciones Unidas, Barcelona, 1969, pp. 317-325.

62. R. Wickbold, The concentration and separation of surface-active agents from surface waters through transport in the gas/water interference. Tenside 5, 61-63 (1971).

Chapter 6

THE VOLUMETRIC ESTIMATION OF ANIONIC SURFACTANTS

Erich Heinerth[*]

Analytical Laboratories
Henkel and Cie, GmbH
Düsseldorf, Federal Republic of Germany

[*]Retired.

221

I. INTRODUCTION

Visual volumetry (titrimetry), one of the simplest and fastest routine methods for the estimation of an extremely wide variety of substances, relies
upon a rapid, stoichiometric reaction between two compounds and an easily
discernible end point.

Normally, cationic and anionic surfactants react stoichiometrically,
in the strict sense of the word, to produce 1:1 salts [11] which are either
insoluble or only sparingly soluble in water but readily soluble in chlorinated
hydrocarbons. The end point of the reaction is not readily observed without
the use of indicators. Many indicators have been proposed (see the exhaustive review by Cross [1]) but alone they do not make the end point sufficiently
distinct [2, 3]; observation of the simultaneous color change of an indicator
and flocculation of the precipitate can sometimes be helpful [4]. Physical
methods for end-point detection have also been suggested, including optical
transmittance [5], nephelometry [6], stalagmometry [7], and nonfaradaic
potentiometry [8], but they are frequently restricted to certain concentrations of accompanying inorganic salts.

The decisive improvement in the visual titrimetric procedure was the
addition of a water-immiscible organic solvent in which a sharp color change
occurs at the end point [9]. This so-called "two-phase titration" may be
performed in acid solution and is further improved by use of a mixture of
two indicators [10].

This chapter deals solely with direct titration methods, although back-
titration methods, such as addition of an excess of cationic surfactant followed by back titration with sodium tetraphenylboron [29] have been reported.

II. TWO-PHASE TITRATION OF
    ANIONIC SURFACTANTS

A. Theoretical Considerations

The color changes which occur in the organic phase at the end of the titration depend upon the nature of the indicator. A basic dye indicator (e.g.,
methylene blue, MB) initially forms a small amount of a colored 1:1 salt
with the anionic surfactant (An^-D^+) which is extracted into the organic phase,
Equation (1). As the mixture is titrated, the cationic surfactant forms a
1:1 salt (An^-Cat^+) with the anionic surfactant and this, also, is extracted
into the organic phase, Equation (2). At the end point, with the free An^-
supply exhausted, Cat^+ displaces the indicator from An^-D^+ and the organic
phase decolorizes as the free indicator returns to the aqueous layer, Equation (3).

$$An^-_{aq} + D^+_{aq} \longrightarrow An^-D^+_{org} \tag{1}$$

$$An^-_{aq} + Cat^+_{aq} \longrightarrow An^-Cat^+_{org} \tag{2}$$

$$An^-D^+_{org} + Cat^+_{aq} \longrightarrow An^-Cat^+_{org} + D^+_{aq} \tag{3}$$

In the alternate procedure, the anion of an acid dye indicator, D^- (e.g., bromophenol blue) remains in the aqueous phase during the titration and thus the organic phase is colorless. At the end of the titration, a small excess of cationic surfactant reacts with D^-, forming D^-Cat^+ which is extracted into the organic phase and thus imparts color to it, Equation (4).

$$An^-_{aq} + Cat^+_{aq} \longrightarrow An^-Cat^+_{org} \tag{2}$$

$$D^-_{aq} + Cat^+_{aq} \longrightarrow D^-Cat^+_{org} \tag{4}$$

Thus basic dye indicators show the end point of the titration by decoloration of the organic phase, and acid dye indicators by the appearance of color therein. These considerations are only valid upon the following assumptions:

1. The 1:1 salt, An^-Cat^+, is quantitatively extracted into the organic phase, leaving no dissociated ions.

2. The salt An^-D^+ is essentially more soluble in the organic phase than in water.

3. The salt An^-D^+ rapidly exchanges the basic dye cation, D^+ for the titrant ion, Cat^+, but not until the free anionic surfactant ions, An^-, have reacted quantitatively.

4. Analogous to 2 and 3, the salt Cat^+D^- is essentially more soluble in the organic phase than in water but is not formed until Cat^+ has reacted quantitatively with An^-.

In practice, although dyestuffs do form 1:1 compounds with surfactants [12, 13] and assumptions 2, 3, and 4 can be fulfilled by selection of the appropriate indicator, varying degrees of departure from assumption 1 are observed.

The natures of the hydrophobic and hydrophilic groups of the surfactants play an important role in the titration, just as they govern the detergent properties. Compounds with short alkyl chains show a stronger hydrophilic nature than do their longer-chain homologues and are very soluble in water. The partition of their salts with cationic surfactants between the phases tends to favor the aqueous phase to an extent which makes the titration unfeasible. In addition, chain branching and the siting of the hydrophile are important [14].

Exact figures for the extractability of the salts formed between cationic and anionic surfactants are not available. However, the differences between the extractabilities of some common types of anionic surfactants with iron(II) chelate cations into chloroform have been reported by Taylor and Waters [15] (see also Chapter 5), who showed that the extraction constants for an alkyl sulfate (C_n) corresponded to that of an alkanesulfonate with a slightly longer chain (C_{n+2}) and an alkylbenzenesulfonate corresponding to $C_{n-2}C_6H_4SO_3H$; thus the influence of the hydrophilic portion of the molecule is apparent.

B. Choice of Titrant

In principle, all strong cationic surfactants with cations of molecular weight around 300 should be suitable (e.g., tetraalkyl ammonium, alkylpyridinium, and benzyltriarylphosphonium salts). However, the titrant must be commercially available in a pure crystalline form and dissolve readily in water to yield a stable, colorless solution. Those most frequently used at present are cetyltrimethylammonium bromide (CTAB), benzyldimethylalkylammonium chloride (BAC), and p-tert-octyl (phenoxyethoxyethyl) dimethylbenzylammonium chloride (Hyamine 1622, Rohm and Haas). The latter is to be preferred [16, 17], having also been shown to be superior to the chlorides of dialkyldimethylammonium, alkylpyridinium, alkylbenzylhydroxyethylammonium, and benzyltriarylphosphonium [18].

Attempts to extend the titration to anionic surfactants with short alkyl chains by using titrants with higher molecular weights were unsuccessful [18].

C. Choice of Indicator

A suitable cationic or anionic indicator should be the salt of a strong inorganic acid or base, respectively, which is appreciably soluble in water, insoluble in hydrocarbons or chlorinated hydrocarbons, and is available commercially in a pure crystalline form; the presence of by-products or impurities tends to obscure the end point. Many indicators have been proposed, reviewed by Cross [1] and Rosen and Goldsmith [3]. Basic dyes are theoretically preferred as indicators since no excess of titrant is required to induce the color change (see Sec. II.A). The frequently performed "Epton titration" uses methylene blue although the acid dyestuff bromophenol blue is often employed, and the mixed indicator (MI) reported by Herring [10] has been widely adopted [19-22]. Mixed indicators (MI) are commonly used in visual titrimetry. They consist of two substances of contrasting colors that make the overall color change more distinctive [23].

In this case the MI is a 1:1 w/w mixture of the anionic dye Disulfine Blue
VN 150 (4',4"-diaminodiethyltriphenylmethane-2,4 disulfonate, CI) and the
cationic dye dimidium bromide (2',7-diamino-9-phenyl-10-methylphenan-
thridinium bromide, actually an abandoned trypanocide). Wang and Panzardi [30] replaced the mixed indicator with the cationic indicator azure A
and the anionic indicator methyl orange, the latter being added just before
the end point. This method has not yet been subjected to the rigorous
testing by any of the major standards associations as have the other methods discussed herein, and hence its reproducibility between laboratories
has not been established.

D. Choice of Second Phase

The organic phase should be water immiscible, colorless, stable, nonviscous, nontoxic, easily obtainable in a pure form, of low volatility, and a
good solvent for the salts formed between the surfactants themselves and
between the surfactants and their counter-ionic dyestuff. Amongst others,
chloroform, dichloromethane, di- and tetrachloroethanes, and hexane have
been proposed; chloroform is most commonly used. Carbon tetrachloride
is not suitable for use with MI [24] but is satisfactory with Wang's mixture [30].

E. The End Point of the Titration

In the case of methylene blue, decoloration of the organic phase marks the
logical end point, but this point is not easily detected in contact with the
colored aqueous phase. Thus Epton [25] selected an end point where both
phases possessed equal color intensities and the same procedure has been
endorsed by ASTM in method D1681-74. The fact that some MB remains
in the organic phase infers that the anionic surfactant has not been quantitatively titrated and hence some form of compensation or correction must
be applied [26].

The use of the mixed indicator provides a different case. At the com-
mencement the organic phase is pink and the aqueous phase is blue due to
dissolved dimidium anionic surfactant salt and disulfine blue, respectively.
As the titration proceeds the organic phase becomes progressively less
pink and turns gray at the end point as a trace of cationic surfactant/disul-
fine blue salt enters it before the last traces of dimidium compounds is
removed. Control of the relative proportions of both dyes result in a
"neutral" gray, but unskilled operators often find the transition more
readily discernible at the point where a faint blue tint appears in the gray
organic layer.

F. Recommended Concentration of the
   Titrant/Titrand*

Standardized procedures, such as ASTM method D1681-74 (1974), DGF
method H-III 10 (1971), and draft ISO recommendation 2271 (1971), pre-
scribe titrant/titrand concentrations in the range 0.004 to 0.005 M. Higher
concentrations are unsuited since their emulsifying powers are too great to
permit adequate separation of the two phases. Lower concentrations down
to 0.001 M may be used and are often preferable to negate interferences by
nonionic surfactants and hydrotropes.

III. SCOPE OF THE METHOD

The two-phase or Epton titration using MB or MI is restricted to the determination of anionic surfactants containing one $-OSO_3^-$ or $-SO_3^-$ group per
mole, typified by:

Sulfates of

    n- and sec-Alkanols
    Ethoxylated alkylphenols
    Ethoxylated alkanols

Sulfonates of

    Alkanes
    Hydroxyalkanes
    Alkenes
    Alkylbenzenes
    Fatty acid/hydroxyethane condensate
    Fatty acid/aminoethane condensate
    Fatty acid esters
    Dialkyl succinate

Frequently MB permits determination of derivatives containing shorter alkyl chains than does MI, such as [18-24]:

Lowest homologs to be determined (recoveries > 95%)

	MB	MI
n-alkyl sulfates	C_8	C_{10}
n-alkanesulfonates	C_{10}	C_{12}
n-alkylbenzenesulfonates	C_6	C_8

---

*"Titrand" is short for "substance to be titrated."

These results were obtained using pure crystalline alkane-1-sulfonates and sulfonates of 1-phenylalkanes. Technical alkanesulfonates are mixtures of mostly sec-alkanesulfonates, and technical (linear) alkylbenzene sulfonates are usually mixtures of holologous 2-, 3-, 4-, etc. phenylalkanes; these may behave somewhat differently than the pure compounds. For that reason it is advisable to standardize the titrant against the type of surfactant to be determined [18, 20]. For comparative studies by different laboratories, standardization against pure sodium lauryl sulfate [16] or sodium tetrapropylenebenzene sulfonic acid methyl ester [20] is recommended.

The $-OSO_3^-$ groups of alkane disulfates react independently of each other and stoichiometrically with Hyamine 1622 in the presence of MB as an indicator, provided that the sulfate groups are separated by not less than four carbon atoms, e.g., salts of decane-1,10-diol disulfate or of 2-octadecyl-1,4-diol disulfate can be determined in this way [26]. MI is not suited to such substances.

Alkanedisulfonates are present in alkanesulfonates produced by sulfoxidation or sulfochlorination of alkanes. In pure form they do not react with Hyamine 1622, but their presence leads to good results in the two-phase titration method; the addition of 2 g of sodium sulfate ensures that correct results are obtained using MI [27].

IV. INTERFERENCES

The two-phase titration of an anionic surfactant using MI is relatively free from interference by most other materials found in admixture. Provided that the sample has been neutralized and any hypochlorite (which destroys the indicator) has been removed with hydrogen peroxide, the inorganic ingredients of detergents, cleansers, scouring powders, etc. (such as phosphates, sulfates, chlorides, borates, perborates, silicates, and carbonates) have no effect.

Organic additives may interfere. Usual ingredients such as EDTA, NTA, urea, nonionic surfactants, alkylolamides, and soap do little to upset the titration but lower alcohols tend to prevent the separation of the phases and are best removed by evaporation. Hydrotropes such as toluene- and xylenesulfonates may be tolerated in quantities up to 15% of the anionic surfactant [28] but cumenesulfonate interferes more strongly. It is recommended that the actual tolerable level of hydrotropes be investigated in each specific case since their influence varies with the type of anionic surfactant being determined [28]. Needless to say, amine oxides interfere by acting as cationic surfactants.

The two-phase titration using MI, being relatively insensitive to the presence of neutral salts, is well-suited to the mutual determination of mixtures of hydrolyzable and nonhydrolyzable surfactants by introduction of an intermediate hydrolysis step with subsequent neutralization. High

excesses of neutral salts may have an effect upon the end point but may be
removed by precipitation with ethanol [17].

V. CONCLUSION

The two-phase titration with a cationic surfactant, preferably Hyamine
1622, in the presence of methylene blue or Herring's mixed indicator (preferred) is a rapid and reliable method for the determination of anionic surfactants as such or as components of consumer products. Approved versions of the methods, together with details of their accuracy and precision,
appear in the following Appendixes.

APPENDIX
Approved Methods for the Determination of Anionic Surfactants by the Two-
phase Titration Methods [16, 17, 19-22]

Mixed indicator	Methylene blue indicator

a. Apparatus

Normal laboratory equipment; 100 ml stoppered measuring cylinder; 25 ml
semimicroburet.

b. Reagents

Water (distilled or deionized); chloroform; ethanol, 95-96% v/v; ethanol,
10% v/v; sulfuric acid, sp gr 1.84; sulfuric acid, 5 N (add 134 ml sulfuric
acid, sp gr 1.84 to about 300 ml water, cool, and dilute to 1 liter); sulfuric
acid, 1 N; sodium hydroxide, 1 N; sodium sulfate, anhydrous; phenolphthalein
solution, 1% in ethanol.

Stock indicator solutions

Mixed indicator	Methylene blue indicator
Dissolve completely 0.5 ± 0.005 g of dimidium bromide (e.g., Burroughs Wellcome Co. Ltd., UK and Gallard-Schlesinger Chemical Manufacturing Comp., USA) and 0.25 ± 0.005 g of Disulfine Blue VN 150 (e.g. ICI, UK, and Gallard Schlesinger, USA) each in 25 ml hot 10% ethanol. Transfer both	Dissolve 0.300 g of methylene blue chloride monohydrate in water, make up to 100 ml and mix thoroughly. To prepare the working indicator solution dilute 10 ml of the stock solution to about 300 ml, add 6.6 g sulfuric acid (sp gr 1.84) and 50 g of sodium sulfate and shake until dissolved. Dilute to

Mixed indicator	Methylene blue indicator
solutions to a 250 ml SVF, dilute to the mark with water and mix thoroughly. The working indicator solution is prepared by diluting 20 ml of the stock solution to 500 ml with water containing 20 ml of 5 N sulfuric acid.	1 liter and mix thoroughly. The solution is 0.000089 M with respect to methylene blue.

Titrant solution

A 0.004 M solution of Hyamine 1622 is prepared by dissolving 1.792 ± 0.001 g of Hyamine 1622 (Rohm and Haas, USA and UK), dried overnight at 105° C, in water and diluting to 1 liter.

c. Determination of Anionic Active Matter

1. Weigh accurately a quantity of sample which contains about 4 meq of anionic active matter with respect to $-OSO_3^-$ and $-SO_3^-$, dissolve it in water. Add a few drops of phenolphthalein indicator and neutralize to a faint pink with 1 N sulfuric acid or sodium hydroxide as required, finishing with the hydroxide.

2. Transfer the neutralized sample to a 1 liter SVF and dilute. Any foam occurring in the neck can be destroyed by addition of a few drops of ethanol. Make up to the mark with water and mix thoroughly.

3.I. Pipette 20 ml of the titrand solution into the 100 ml measuring cylinder and add

15 ml chloroform 10 ml water and 10 ml MI solution	15 ml chloroform and 25 ml MB solution
Add 5 ml titrant from the buret, shake well for 30 sec and allow the layers to separate; the lower layer will be pink.	Add 5 ml titrant from the buret, shake well for 30 sec and allow the layers to separate; the lower layer will be blue.

3.II. Continue titrating with the 0.004 M titrant, initially in 1 ml increments. After each addition shake vigorously and permit the layers to separate. As the end point is approached, the emulsion formed by shaking

(continued)

Mixed indicator	Methylene blue indicator

tends to break easily. Continue the titration dropwise until the end point is reached, i.e., when

the pink color is completely discharged from the chloroform layer which is then a faint gray-blue; with excess titrant the lower layer is blue.	the color is the same in the two layers, viewed in reflected light from a white background after one minute's standing; with excess titrant the chloroform layer is less blue than the aqueous layer.

d. Calculation

$$\text{Anionic surfactant} = \frac{VM}{1000} \times MW \times \frac{100}{S} \text{ wt \%}$$

$$= \frac{V \times M \times MW}{10S} \text{ wt \%}$$

where

V = volume of titrant consumed in ml
M = molarity of the titrant
MW = molecular weight of titrand
S = weight of sample in the aliquot taken, in gram

e. Correction

No correction required	Certain amounts of methylene blue/titrand salt remain in the chloroform layer at the end point. Add to the volume of titrant consumed the volume given by $$\frac{15}{45 + V} \times \frac{0.000089}{M} \times 25 \text{ ml}$$ This correction is almost negligible with 0.004 M titrant, but is usually appreciable (0.5–0.6 ml) with 0.001 M titrant.

Mixed indicator	Methylene blue indicator

f. Standardization of Titrant

For the determination of unidentified or uncharacterized surfactants, the titrant can be standardized against pure sodium lauryl sulfate [20] (BDH, UK or Gallard-Schlesinger, USA) or distilled tetrapropylenebenzenesulfonic acid methyl ester [21]. In all other cases, standardization with the surfactant to be determined is recommended.

g. Remarks

The monosulfonate content of mixed alkanemono- and disulfonates may be determined by addition of 2 g sodium sulfate to the titration mixture [27].

h. Repeatability

The maximum difference between two determinations should not exceed 1.5% of the average value.

i. Reproducibility

The difference between the results of two different laboratories should not exceed 3% of the average value.

REFERENCES

1. J. T. Cross, in Cationic Surfactants (E. Jungermann, ed.), Marcel Dekker, New York, 1970, pp. 452-476.
2. K. Toei and K. Kawada, Fundamental studies on colloid titrations I. Japan Analyst 21, 1510-1515 (1972).
3. M. J. Rosen and H. A. Goldsmith, Systematic Analysis of Surface-Active Agents, 2nd ed., Wiley-Interscience, New York, 1972, pp. 419-430.
4. R. Bennewitz and K. Fiedler, The determination of anionic and cationic surfactants by flocculation-antagonist titration. Tenside 2, 337-340 (1965).
5. P. A. Rodriguez, Automated titration of anionic surfactants. J. Am. Oil Chemists' Soc. 51, 277A (1974).
6. M. Hellsten, Titration of Anionic Surfactants with Cationic Surfactants. Instrumental method for the end-point determination. Proc. 5th Intern. Congress on Surface-Active Substances, Barcelona 1968, Vol. 1, Ediciones Unidas S.A., Barcelona, 1969, pp. 291-298.

7. T. Kambara and T. Kiba, Stalagmometric titrations. Talanta 19, 399-406 (1972).

8. T. Kiba and T. Kambara, Nonfaradaic potentiometric titration of ionic surfactants. J. Electroanal. Chem. Interfacial Electrochem. 44, 129-135 (1973).

9. S. R. Epton, A new method for the rapid titrimetric analysis of sodium alkyl sulphates and related compounds. Trans. Faraday Soc. 44, 226-230 (1948).

10. D. E. Herring, Two-phase titrations. Lab. Pract. 11, 113-115 (1962).

11. M. Mitsuishi and M. Hashizume, On the interaction of sodium alkyl sulfate with alkylpyridinium chloride and alkyltrimethylammonium bromide in solution. Bull. Chem. Soc. Japan 46, 1946-1948 (1973).

12. P. Mukerjee and K. J. Mysels, A re-evaluation of the spectral-change method of determining critical micelle concentrations. J. Am. Chem. Soc. 77, 2937-2943 (1955).

13. H. K. Biswas and B. M. Mandal, Extraction of anions into chloroform by surfactant cations. Relevance to dye extraction method of analysis of long chain amines. Anal. Chem. 44, 1636-1640 (1972).

14. J. J. Lin, B. M. Moudgil, and P. Somasundaran, Estimation of the effective number of $-CH_2$-groups in long-chain surface active agents. Colloid Polymer Sci. 252, 407-414 (1974).

15. C. G. Taylor and J. Waters, Ion-association compounds of anionic surfactants with iron(II) chelates. Part I. Extraction constants. Anal. Chim. Acta 69, 363-371 (1974).

16. V. W. Reid, G. F. Longman, and E. Heinerth, Determination of anionic-active detergents by two-phase titration. Tenside 4, 292-304 (1967).

17. V. W. Reid, G. F. Longman, and E. Heinerth, Determination of anionic-active detergents by two-phase titration (II). Tenside 5, 90-96 (1968).

18. E. Heinerth, The two-phase titration of anionic and cationic surfactants. Fette, Seifen, Anstrichmittel 72, 385-388 (1970).

19. Commission Internationale d'Analyses du Comité International des Dérivés Tensio-Actifs, Paris, Determination of Anionic-Active Detergents, Direct Two-phase Titration Procedure, Doc. CIA 2-70 E, 1970.

20. Deutsche Gesellschaft für Fettwissenschaft Einheitsmethoden zur Untersuchung von Fetten, Fettprodukten und verwandten Stoffen. Anionic surfactants (two-phase titration). Fette, Seifen, Anstrichmittel 73, 683-684 (1971).

21. Draft ISO Recommendation 2271, Determination of Anionic-Active Matter, Direct Two-Phase Titration Procedure, International Organization for Standardization, Geneva, Switzerland, 1971.

22. ASTM D 3049-72 T, Test for Synthetic Anionic Active Ingredient in Detergents by Cationic Titration Procedure, Part 30. 1974 Annual Book of ASTM Standards, American Society for Testing and Materials, Philadelphia, 1974.

23. C. L. Hilton, in Encyclopedia of Industrial Chemical Analysis (F. D. Snell and C. L. Hilton, eds.), Vol. 2, Wiley-Interscience, New York, 1966, p. 248.

24. D. Owen and A. T. Pugh, Determination of the "potential surface-active hydroxyl content" of fatty alcohols and their ethoxylates. Analyst 100, 269-274 (1975).

25. S. R. Epton, A rapid method of analysis of certain surface-active agents. Nature 160, 795-796 (1947).

26. H. Y. Lew, Some new developments in surfactant analysis. J. Am. Oil Chemists' Soc. 49, 665-670 (1972).

27. R. Wickbold, Progress in the examination of alkane sulfonates. Tenside 8, 130-134 (1971).

28. N. A. Puttnam and P. Platt, The extent of interference of hydrotropes in the direct two-phase titration procedure (Draft ISO Recommendation 2271). Proc. 6th Intern. Congress on Surface-Active Substances, Zürich 1972, Vol. I, Carl Hanser Verlag, Munchen, 1973, pp. 433-438.

29. L. K. Wang, J. Y. Yang, and M. H. Wang, Proposed method for the analysis of anionic surfactants. J. Am. Water Works Assoc. 67, 6-8 (1976).

30. L. K. Wang and P. J. Panzardi, Determination of anionic surfactants with azure A and quaternary ammonium salt. Two-phase titration. Anal. Chem. 47, 1472 (1975).

AUTHOR INDEX

Numbers in brackets are reference numbers and indicate that an author's work is referred to although his name is not cited in the text. Underlined numbers give the page on which the complete reference is listed.

SUBJECT INDEX

247